THE FOUNDATIONS OF
GEOMETRY

THE FOUNDATIONS OF
GEOMETRY

David Hilbert, Ph.D.

Professor of Mathematics
University of Gottingen

Translated By

E.J. Townsend, Ph.D.
University of Illinois

MJP PUBLISHERS

Reprint of 1902 Edition
MJP first reprint : 2008

ISBN 81-8094-053-5   **MJP PUBLISHERS**
© Publishers, 2008   47, Nallathambi Street
 Triplicane
Printed and bound in India   Chennai 600 005
MJP 045

PREFACE

The material contained in the following translation was given in substance by Professor Hilbert as a course of lectures on euclidean geometry at the University of Gottingen during the winter semester of 1898–1899. The results of his investigation were re-arranged and put into the form in which they appear here as a memorial address published in connection with the celebration at the unveiling of the Gauss-Weber monument at Gottingen, in June, 1899. In the French edition, which appeared soon after, Professor Hilbert made some additions, particularly in the concluding remarks, where he gave an account of the results of a recent investigation made by Dr. Dehn. These additions have been incorporated in the following translation.

As a basis for the analysis of our intuition of space, Professor Hilbert commences his discussion by considering three systems of things which he calls points, straight lines, and planes, and sets up a system of axioms connecting these elements in their mutual relations. The purpose of his investigations is to discuss systematically the relations of these axioms to one another and also the bearing of each upon the logical development of euclidean geometry. Among the important results obtained, the following are worthy of special mention:

1. The mutual independence and also the compatibility of the given system of axioms is fully discussed by the aid of various new systems of geometry which are introduced.

2. The most important propositions of euclidean geometry are demonstrated in such a manner as to show precisely what axioms underlie and make possible the demonstration.

3. The axioms of congruence are introduced and made the basis of the definition of geometric displacement.

4. The significance of several of the most important axioms and theorems in the development of the euclidean geometry is clearly shown; for example, it is shown that the whole of the euclidean geometry may be developed without the use of the axiom of continuity; the significance of Desargues's theorem, as a condition that a given plane geometry may be regarded as a part of a geometry of space, is made apparent, etc.

5. A variety of algebras of segments are introduced in accordance with the laws of arithmetic.

This development and discussion of the foundation principles of geometry is not only of mathematical but of pedagogical importance. Hoping that through an English edition these important results of Professor Hilbert's investigation may be made more accessible to English speaking students and teachers of geometry, I have undertaken, with his permission, this translation. In its preparation, I have had the assistance of many valuable suggestions from Professor Osgood of Harvard, Professor Moore of Chicago, and Professor Halsted of Texas. I am also under obligations to Mr. Henry Coar and Mr. Arthur Bell for reading the proof.

E J Townsend
University of Illinois

CONTENTS

INTRODUCTION

"All human knowledge begins with intuitions, thence passes to concepts and ends with ideas."

> Kant, Kritik der reinen Vernunft,
> Elementariehre, Part 2, Sec. 2.

Geometry, like arithmetic, requires for its logical development only a small number of simple, fundamental principles. These fundamental principles are called the axioms of geometry. The choice of the axioms and the investigation of their relations to one another is a problem which, since the time of Euclid, has been discussed in numerous excellent memoirs to be found in the mathematical literature.[1] This problem is tantamount to the logical analysis of our intuition of space.

The following investigation is a new attempt to choose for geometry a *simple* and *complete* set of *independent* axioms and to deduce from these the most important geometrical theorems in such a manner as to bring out as clearly as possible the significance of the different groups of axioms and the scope of the conclusions to be derived from the individual axioms.

1

THE FIVE GROUPS OF AXIOMS

§1. THE ELEMENTS OF GEOMETRY AND THE FIVE GROUPS OF AXIOMS

Let us consider three distinct systems of things. The things composing the first system, we will call *points* and designate them by the letters *A, B, C,.* . . ; those of the second, we will call *straight lines* and designate them by the letters *a, b, c,.* . . ; and those of the third system, we will call *planes* and designate them by the Greek letters ", $\alpha, \beta, \gamma,$ ... The points are called the *elements of linear geometry*; the points and straight lines, the *elements of plane geometry*; and the points, lines, and planes, the *elements of the geometry of space* or the *elements of space.*

We think of these points, straight lines, and planes as having certain mutual relations, which we indicate by means of such words as "are situated," "between," "parallel," "congruent," "continuous," etc. The complete and exact description of these relations follows as a consequence of the *axioms of geometry*. These axioms may be arranged in five groups. Each of these groups expresses, by itself, certain related fundamental facts of our intuition. We will name these groups as follows:

    I.   1–7. Axioms of *connection.*

    II. 1–5. Axioms of *order.*

III. Axiom of *parallels* (Euclid's axiom).

IV. 1–6. Axioms of *congruence*.

V. Axiom of *continuity* (Archimedes's axiom).

§2. GROUP I: AXIOMS OF CONNECTION

The axioms of this group establish a connection between the concepts indicated above; namely, points, straight lines, and planes. These axioms are as follows:

I, 1.   *Two distinct points A and B always completely determine a straight line a. We write AB = a or BA = a.*

Instead of "determine," we may also employ other forms of expression; for example, we may say A "lies upon" a, A "is a point of" a, a "goes through" A "and through" B, a "joins" A "and" or "with" B, etc. If A lies upon a and at the same time upon another straight line b, we make use also of the expression: "The straight lines" a "and" b "have the point A in common," etc.

I, 2.   *Any two distinct points of a straight line completely determine that line; that is, if AB = a and AC = a, where B ≠ C, then is also BC = a.*

I, 3.   *Three points A, B, C not situated in the same straight line always completely determine a plane α. We write ABC = a.*

We employ also the expressions: *A, B, C,* "lie in" α; *A, B, C* "are points of" α, etc.

I, 4.   *Any three points A, B, C of a plane α", which do not lie in the same straight line, completely determine that plane.*

I, 5.   *If two points A, B of a straight line a lie in a plane α, then every point of a lies in α.*

In this case we say: "The straight line *a* lies in the plane *α*," etc.

I, 6.   *If two planes α, β have a point A in common, then they have at least a second point B in common.*

I, 7.   *Upon every straight line there exist at least two points, in every plane at least three points not lying in the same straight line, and in space there exist at least four points not lying in a plane.*

Axioms I, 1–2 contain statements concerning points and straight lines only; that is, concerning the elements of plane geometry. We will call them, therefore, the *plane axioms of group I*, in order to distinguish them from the axioms I, 3–7, which we will designate briefly as the *space axioms* of this group.

Of the theorems which follow from the axioms I, 3–7, we shall mention only the following:

Theorem 1

Two straight lines of a plane have either one point or no point in common; two planes have no point in common or a straight line in common; a plane and a straight line not lying in it have no point or one point in common.

Theorem 2

Through a straight line and a point not lying in it, or through two distinct straight lines having a common point, one and only one plane may be made to pass.

3. GROUP II: AXIOMS OF ORDER.[2]

The axioms of this group define the idea expressed by the word "between," and make possible, upon the basis of this idea, an *order of sequence* of the points upon a straight line, in a plane, and in space. The points of a straight line have a certain relation to one another which the word "between" serves to describe. The axioms of this group are as follows:

II, 1.   *If A, B, C are points of a straight line and B lies between A and C, then B lies also between C and A.*

Figure 1

II, 2.   *If A and C are two points of a straight line, then there exists at least one point B lying between A and C and at least one point D so situated that C lies between A and D.*

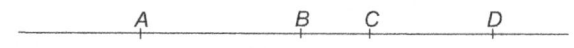

Figure 2

II, 3.   *Of any three points situated on a straight line, there is always one and only one which lies between the other two.*

II, 4.   *Any four points A, B, C, D of a straight line can always be so arranged that B shall lie between A and C and also between A and D, and, furthermore, that C shall lie between A and D and also between B and D.*

Definition

We will call the system of two points *A* and *B*, lying upon a straight line, a *segment* and denote it by *AB* or *BA*. The points lying between *A* and *B* are called the *points of the segment AB* or the *points lying within the segment AB*. All other points of the straight line are referred to as the *points lying outside the segment AB*. The points *A* and *B* are called the *extremities* of the segment *AB*.

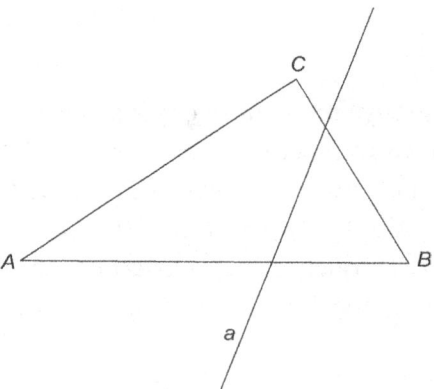

Figure 3

II, 5. *Let A, B, C be three points not lying in the same straight line and let a be a straight line lying in the plane ABC and not passing through any of the points A, B, C. Then, if the straight line a passes through a point of the segment AB, it will also pass through either a point of the segment BC or a point of the segment AC.* Axioms II, 1–4 contain statements concerning the points of a straight line only, and, hence, we will call them the *linear axioms of group II*. Axiom II, 5 relates to the elements of plane geometry and, consequently, shall be called the *plane axiom of group II*.

§4. CONSEQUENCES OF THE AXIOMS OF CONNECTION AND ORDER

By the aid of the four linear axioms II, 1–4, we can easily deduce the following theorems:

Theorem 3

Between any two points of a straight line, there always exists an unlimited number of points.

Theorem 4

If we have given any finite number of points situated upon a straight line, we can always arrange them in a sequence $A, B, C, D, E,..., K$ so that B shall lie between A and $C, D, E,..., K; C$ between A, B and $D, E,..., K; D$ between A, B, C and $E,... K$, etc. Aside from this order of sequence, there exists but one other possessing this property namely, the reverse order $K,..., E, D, C, B, A$.

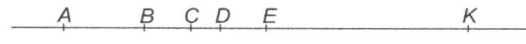

Figure 4

Theorem 5

Every straight line a, which lies in a plane α, divides the remaining points of this plane into two regions having the following properties: Every point A of the one region determines with each point B of the other region a segment AB containing a point of the straight line a. On the other hand, any two points A, A' of the same region determine a segment AA' containing no point of a.

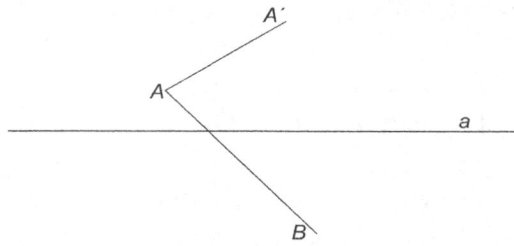

Figure 5

If A, A', O, B are four points of a straight line a, where O lies between A and B but not between A and A', then we may say: The points A, A' are situated *on the line a upon one and the same side of the point O*, and the points A, B are situated *on the straight line a upon different sides of the point O*.

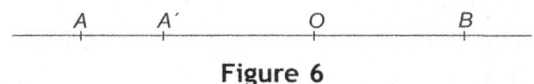

Figure 6

All of the points of a which lie upon the same side of O, when taken together, are called the *half-ray* emanating from O. Hence, each point of a straight line divides it into two half-rays.

Making use of the notation of theorem 5, we say: The points A, A' lie *in the plane α upon one and the same side of the straight line a*, and the points A, B lie *in the plane α upon different sides of the straight line a*.

Definitions

A system of segments AB, BC, CD, ..., KL is called a *broken line* joining A with L and is designated, briefly, as the broken line $ABCDE ... KL$. The points lying within the segments AB, BC, CD, ..., KL, as also the points A, B, C, D, ..., K, L, are called *the points of the broken line*. In particular, if the point A coincides with *L*, the broken line is called a *polygon* and is designated as the polygon *ABCD* ...*K*. The segments *AB, BC, CD,* ..., *KA* are called the *sides of the polygon* and the points *A, B, C, D,* ..., *K* the *vertices*. Polygons having 3, 4, 5, ... , *n* vertices are called, respectively, *t*riangles, *q*uadrangles, *p*entagons, . . . , *n*-gons. If the vertices of a polygon are all distinct and none of them lie within the segments composing the sides of the polygon, and, furthermore, if no two sides have a point in common, then the polygon is called a *simple polygon*.

With the aid of theorem 5, we may now obtain, without serious difficulty, the following theorems:

Theorem 6

Every simple polygon, whose vertices all lie in a plane α, divides the points of this plane, not belonging to the broken line constituting the sides of the polygon, into two regions, an interior and an exterior, having the following properties: If *A* is a point of the interior region (interior point) and *B* a point of the exterior region (exterior point), then any broken line joining *A* and *B* must have at least one point in common with the polygon. If, on the other hand, *A, A'* are two points of the interior and *B, B'* two points of the exterior region, then there are always broken lines to be found joining *A* with *A'* and *B* with *B'* without having a point in common with the polygon. There exist straight lines in the plane α which lie entirely outside of the given polygon, but there are none which lie entirely within it.

Theorem 7

Every plane α divides the remaining points of space into two regions having the following properties: Every point *A* of the one region determines with each point *B* of the other region a segment *AH*, within which lies a point of α. On the other hand, any two points A,A' lying within the same region determine a segment AA' containing no point of α.

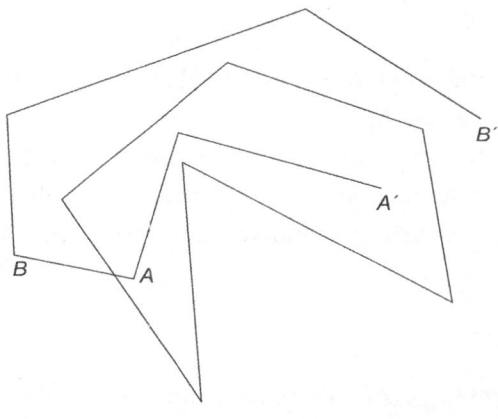

Figure 7

Making use of the notation of theorem 7, we may now say: The points *A, A'* are situated in space *upon one and the same side of the plane* α, and the points *A, B* are situated in space *upon different sides of the plane* α.

Theorem 7 gives us the most important facts relating to the order of sequence of the elements of space. These facts are the results, exclusively, of the axioms already considered, and, hence, no new space axioms are required in group II.

§5. GROUP III: AXIOM OF PARALLELS (EUCLID'S AXIOM)

The introduction of this axiom simplifies greatly the fundamental principles of geometry and facilitates in no small degree its development. This axiom may be expressed as follows:

III. *In a plane α there can be drawn through any point A, lying outside of a straight line a, one and only one straight line which does not intersect the line a. This straight line is called the parallel to a through the given point A.*

This statement of the axiom of parallels contains two assertions. The first of these is that, in the plane α, there is always a straight line passing through A which does not intersect the given line a. The second states that only one such line is possible. The latter of these statements is the essential one, and it may also be expressed as follows:

Theorem 8

If two straight lines a, b of a plane do not meet a third straight line c of the same plane, then they do not meet each other.

For, if a, b had a point A in common, there would then exist in the same plane with c two straight lines a and b each passing through the point A and not meeting the straight line c. This condition of affairs is, however, contradictory to the second assertion contained in the axiom of parallels as originally stated. Conversely, the second part of the axiom of parallels, in its original form, follows as a consequence of theorem 8.

The axiom of parallels is a *plane axiom*.

§6. GROUP IV: AXIOMS OF CONGRUENCE

The axioms of this group define the idea of congruence or displacement. Segments stand in a certain relation to one another which is described by the word "*congruent*."

IV, I. *If A, B are two points on a straight line a, and if A´ is a point upon the same or another straight line a´, then, upon a given side of A´ on the straight line a´, we can always find one and only one point B´ so that the segment AB (or BA) is congruent to the segment A´B´. We indicate this relation by writing*

$$AB \equiv A´B´$$

Every segment is congruent to itself; that is, we always have

$$AB \equiv AB$$

We can state the above axiom briefly by saying that every segment can be *laid off* upon a given side of a given point of a given straight line in one and and only one way.

IV, 2. *If a segment AB is congruent to the segment A´B´ and also to the segment A´´B´´, then the segment A´B´ is congruent to the segment A´´B´´; that is, if* $AB \equiv A´B$ *and AB* $\equiv A´´B´´$, *then A´B´* $\equiv A´´B´´$.

IV, 3. *Let AB and BC be two segments of a straight line a which have no points in common aside from the point B, and, furthermore, let A´B´ and B´C´ be two segments of the same or of another straight line a´ having, likewise, no point other than B´ in common. Then, if AB* $\equiv A´B´$ *and BC* $\equiv B´C´$, *we have AC* $\equiv A´C´$.

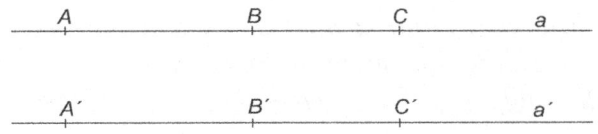

Figure 8

Definitions

Let α be any arbitrary plane and h, k any two distinct half-rays lying in α and emanating from the point O so as to form a part of two different straight lines. We call the system formed by these two half-rays h, k an *angle* and represent it by the symbol $\angle(h, k)$ or $\angle(k, h)$. From axioms II, 1–5, it follows readily that the half-rays h and k, taken together with the point O, divide the remaining points of the plane α into two regions having the following property: If A is a point of one region and B a point of the other, then every broken line joining A and B either passes through O or has a point in common with one of the half-rays h, k. If, however, A, A' both lie within the same region, then it is always possible to join these two points by a broken line which neither passes through O nor has a point in common with either of the half-rays h, k. One of these two regions is distinguished from the other in that the segment joining any two points of this region lies entirely within the region. The region so characterised is called the *interior of the angle* (h, k). To distinguish the other region from this, we call it the *exterior of the angle* (h, k). The half rays h and k are called the sides of the angle, and the point O is called the *vertex of the angle*.

IV, 4. *Let an angle (h, k) be given in the plane α and let a straight line a' be given in a plane α'. Suppose also that, in the plane α, a definite side of the straight line a' be assigned. Denote by h' a half-ray of the straight line a' emanating from a point O' of this line. Then in the plane α' there is one and only one half-ray k' such that the angle (h, k), or (k, h), is congruent to the angle (h',k') and at the same time all interior points of the angle (h',k') lie upon the given side of a'. We express this relation by means of the notation*

$$\angle(h, k) \equiv \angle(h', k')$$

Every angle is congruent to itself; that is,

$$\angle(h, k) \equiv \angle(h, k)$$

or

$$\angle(h, k) \equiv \angle(k, h)$$

We say, briefly, that every angle in a given plane can be *laid off* upon a given side of a given half-ray in one and only one way.

IV, 5. *If the angle (h, k) is congruent to the angle (h', k') and to the angle (h'', k''); then the angle (h', k') is congruent to the angle (h'', k''); that is to say, if $\angle(h, k) \equiv \angle(h', k')$ and $\angle(h, k) \equiv \angle(h'', k'')$, then $\angle(h', k') \equiv \angle(h'', k'')$.*

Suppose we have given a triangle ABC. Denote by h, k the two half-rays emanating from A and passing respectively through B and C. The angle (h, k) is then said to be the angle included by the sides AB and AC, or the one opposite to the side BC in the triangle ABC. It contains all of the interior points of the triangle ABC and is represented by the symbol $\angle BAC$, or by $\angle A$.

IV, 6. *If, in the two triangles ABC and $A'B'C'$ the congruences*

$$AB \equiv A'B', \quad AC \equiv A'C', \quad \angle BAC \equiv \angle B'A'C'$$

hold, then the congruences

$$\angle ABC \equiv \angle A'B'C' \text{ and } \angle ACB = \angle A'C'B'$$

also hold.

Axioms IV, 1–3 contain statements concerning the congruence of segments of a straight line only. They may, therefore, be called the *linear* axioms of group IV. Axioms IV, 4, 5 contain statements relating to the congruence of angles. Axiom IV, 6 gives the connection between the congruence of segments and the congruence of angles. Axioms IV, 4–6 contain statements regarding the elements of plane geometry and may be called the *plane* axioms of group IV.

§7. CONSEQUENCES OF THE AXIOMS OF CONGRUENCE

Suppose the segment AB is congruent to the segment $A'B'$. Since, according to axiom IV, 1, the segment AB is congruent to itself, it follows from axiom IV, 2 that $A'B'$ is congruent to AB; that is to say, if $AB \equiv A'B'$, then $A'B' \equiv AB$. We say, then, that the two segments are congruent to one another.

Let $A, B, C, D,..., K, L$ and $A', B', C', D',..., K', L'$ be two series of points on the straight lines a and a', respectively, so that all the corresponding segments AB and $A'B'$, AC and $A'C'$, BC and $B'C'$,..., KL and $K'L'$ are respectively congruent, then *the two series of points are said to be congruent to one another*. A and A', B and B',..., L and L' are called *corresponding points* of the two congruent series of points.

From the linear axioms IV, 1–3, we can easily deduce the following theorems:

Theorem 9

If the first of two congruent series of points $A, B, C, D,..., K, L$ and $A', B', C', D',..., K', L'$ is so arranged that B lies between A and $C, D,..., K, L$, and C between A, B and $D,..., K, L$, etc., then the points $A', B', C', D',..., K', L'$ of the second series are arranged in a similar way; that is to say, B' lies between A' and C', $D',..., K', L'$, and C' lies between A', B' and $D',..., K', L'$, etc.

Let the angle (h, k) be congruent to the angle (h',k'). Since, according to axiom IV, 4, the angle (h, k) is congruent to itself, it follows from axiom IV, 5 that the angle (h',k') is congruent to the angle (h, k). We say, then, that the angles (h, k) and (h',k') are *congruent to one another.*

Definitions

Two angles having the same vertex and one side in common, while the sides not common form a straight line, are called *supplementary angles*. Two angles having a common vertex and whose sides form straight lines are called *vertical angles*. An angle which is congruent to its supplementary angle is called a *right angle*.

Two triangles ABC and $A'B'C'$ are said to be *congruent* to one another when all of the following congruences are fulfilled:

$$AB \equiv A'B', \ AC \equiv A'C', \ BC \equiv B'C',$$
$$\angle A \equiv \angle A', \ \angle B \equiv \angle B', \ \angle C \equiv \angle C'.$$

Theorem 10

(First theorem of congruence for triangles). If, for the two triangles ABC and $A'B'C'$, the congruences

$$AB \equiv A'B', \ AC \equiv A'C', \angle A \equiv \angle A'$$

hold, then the two triangles are congruent to each other.

Proof From axiom IV, 6, it follows that the two congruences

$$\angle B \equiv \angle B' \text{ and } \angle C \equiv \angle C'$$

are fulfilled, and it is, therefore, sufficient to show that the two sides BC and $B'C'$ are congruent. We will assume the contrary to be true, namely, that BC and $B'C'$ are not congruent, and show that this leads to a contradiction. We take upon $B'C'$ a point D' such that $BC \equiv B'D'$. The two triangles ABC and $A'B'D'$ have, then, two sides and the included angle of the one agreeing, respectively, to two sides and the included angle of the other. It follows from axiom IV, 6 that the two angles BAC and $B'A'D'$ are also congruent to each other. Consequently, by aid of axiom IV, 5, the two angles $B'A'C'$ and $B'A'D'$ must be congruent.

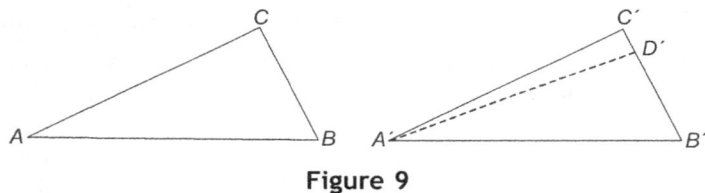

Figure 9

This, however, is impossible, since, by axiom IV, 4, an angle can be laid off in one and only one way on a given side of a given half-ray of a plane. From this contradiction the theorem follows.

We can also easily demonstrate the following theorem:

Theorem 11   (Second Theorem of Congruence for triangles)

If in any two triangles one side and the two adjacent angles are respectively congruent, the triangles are congruent.

We are now in a position to demonstrate the following important proposition.

Theorem 12

If two angles ABC and $A'B'C'$ are congruent to each other, their supplementary angles CBD and $C'B'D'$ are also congruent.

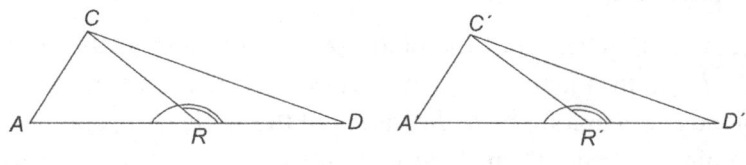

Figure 10

Proof   Take the points A', C', D' upon the sides passing through B' in such a way that

$$A'B' \equiv AB, \ C'B' \equiv CB, \ D'B' \equiv DB$$

Then, in the two triangles ABC and $A'B'C'$, the sides AB and BC are respectively congruent to $A'B'$ and $C'B'$. Moreover, since the angles included by these sides are congruent to each other by hypothesis, it follows from theorem 10 that these triangles are congruent; that is to say, we have the congruences

$$AC \equiv A'C, \angle BAC \equiv \angle B'A'C'$$

On the other hand, since by axiom IV, 3 the segments AD and $A'D'$ are congruent to each other, it follows again from theorem 10 that the triangles CAD and $C'A'D'$ are congruent, and, consequently, we have the congruences:

$$CD \equiv C'D', \angle ADC \equiv \angle A'D'C'$$

From these congruences and the consideration of the triangles BCD and $B'C'D'$, it follows by virtue of axiom IV, 6 that the angles CBD and $C'B'D'$ are congruent.

As an immediate consequence of theorem 12, we have a similar theorem concerning the congruence of vertical angles.

Theorem 13

Let the angle (h, k) of the plane α be congruent to the angle (h', k') of the plane α', and, furthermore, let l be a half-ray in the plane α emanating from the vertex of the angle (h, k) and lying within this angle. Then, there always exists in the plane α' a half-ray l' emanating from the vertex of the angle (h', k') and lying within this angle so that we have

$$\angle(h, l) \equiv \angle(h',l'), \angle(k, l) \equiv \angle(k',l')$$

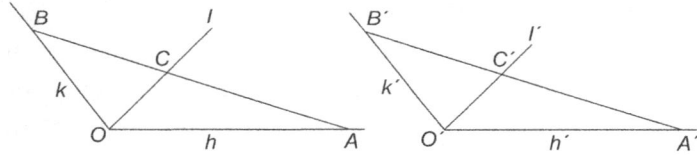

Figure 11

Proof   We will represent the vertices of the angles (h, k) and (h', k') by O and O', respectively, and so select upon the sides h, k, h', k' the points A, B, A', B' that the congruences

$$OA \equiv O'A', \ OB \equiv O'B'$$

are fulfilled. Because of the congruence of the triangles OAB and $O'A'B'$, we have at once

$$AB \equiv A'B', \angle OAB \equiv O'A'B', \angle OBA \equiv \angle O'B'A'$$

Let the straight line AB intersect l in C. Take the point C' upon the segment $A'B'$ so that $A'C' \equiv AC$. Then, $O'C'$ is the required half-ray. In fact, it follows directly from these congruences, by aid of axiom IV, 3, that $BC \equiv B'C'$. Furthermore, the triangles OAC and $O'A'C'$ are congruent to each other, and the same is true also of the triangles OCB and $O'B'C'$. With this our proposition is demonstrated.

In a similar manner, we obtain the following proposition.

Theorem 14

Let h, k, l and h', k', l' be two sets of three half-rays, where those of each set emanate from the same point and lie in the same plane. Then, if the congruences

$$\angle(h, l) \equiv \angle(h', l'), \angle(k, l) \equiv \angle(k', l')$$

are fulfilled, the following congruence is also valid; viz.:

$$\angle(h, k) \equiv \angle(h', k').$$

By aid of theorems 12 and 13, it is possible to deduce the following simple theorem, which Euclid held–although it seems to me wrongly–to be an axiom.

Theorem 15

All right angles are congruent to one another.

Proof Let the angle BAD be congruent to its supplementary angle CAD, and, likewise, let the angle $B'A'D'$ be congruent to its supplementary angle $C'A'D'$. Hence the angles BAD, CAD, $B'A'D'$, and $C'A'D'$ are all right angles. We will assume that the contrary of our proposition is true, namely, that the right angle $B'A'D'$ is not congruent to the right angle BAD, and will show that this assumption leads to a contradiction. We lay off the angle $B'A'D'$ upon the half-ray AB in such a manner that the side AD'' arising from this operation falls either within the angle BAD or within the angle CAD. Suppose, for example, the first of these possibilities to be true. Because of the congruence of the angles $B'A'D'$ and BAD'', it follows from theorem 12 that angle $C'A'D'$ is congruent to angle CAD'', and, as the angles $B'A'D'$. and $C'A'D'$ are congruent to each other, then, by IV, 5, the angle BAD'' must be congruent to CAD''.

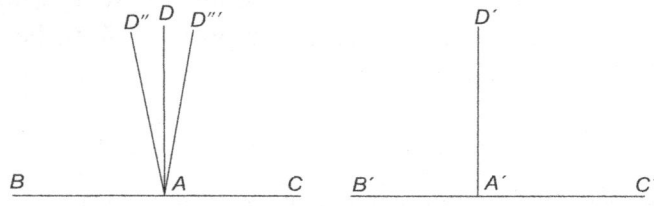

Figure 12

Furthermore, since the angle BAD is congruent to the angle CAD, it is possible, by theorem 13, to find within the angle CAD a half-ray AD''' emanating from A, so that the angle BAD'' will be congruent to the angle CAD''', and also the angle DAD'' will be congruent to the angle DAD'''. The angle BAD'' was shown to be congruent to the angle CAD'' and, hence, by axiom IV, 5, the angle CAD'', is congruent to the angle CAD'''. This, however, is not possible; for, according to axiom IV, 4, an angle can be laid off in a plane upon a given side of a given half-ray in only one way. With this our proposition is demonstrated. We can now introduce, in accordance with common usage, the terms "*acute angle*" and "*obtuse angle.*"

The theorem relating to the congruence of the base angles A and B of an equilateral triangle ABC follows immediately by the application of axiom IV, 6 to the triangles ABC and BAC. By aid of this theorem, in addition to theorem 14, we can easily demonstrate the following proposition.

Theorem 16 (Third Theorem of Congruence for Triangles)

If two triangles have the three sides of one congruent respectively to the corresponding three sides of the other, the triangles are congruent.

Any finite number of points is called a *figure*. If all of the points lie in a plane, the figure is called a *plane figure*.

Two figures are said to be *congruent* if their points can be arranged in a one-to-one correspondence so that the corresponding segments and the corresponding angles of the two figures are in every case congruent to each other.

Congruent figures have, as may be seen from theorems 9 and 12, the following properties: Three points of a figure lying in a straight line are likewise in a straight line in every figure congruent to it. In congruent figures, the arrangement of the points in corresponding planes with respect to corresponding lines is always the same. The same is true of the sequence of corresponding points situated on corresponding lines.

The most general theorems relating to congruences in a plane and in space may be expressed as follows:

Theorem 17

If $(A, B, C, \ldots)$ and $(A', B', C', \ldots)$ are congruent plane figures and P is a point in the plane of the first, then it is always possible to find a point P in the plane of the second figure so that $(A, B, C, \ldots, P)$ and $(A', B', C', \ldots P')$ shall likewise be congruent figures. If the two figures have at least three points not lying in a straight line, then the selection of P' can be made in only one way.

Theorem 18

If $(A, B, C, \ldots)$ and $(A', B', C', \ldots)$ are congruent figures and P represents any arbitrary point, then there can always be found a point P' so that the two figures $(A, B, C, \ldots, P)$ and $(A', B', C', \ldots, P')$ shall likewise be congruent. If the figure $(A, B, C, \ldots, P)$ contains at least four points not lying in the same plane, then the determination of P' can be made in but one way.

This theorem contains an important result; namely, that all the facts concerning space which have reference to congruence, that is to say, to displacements in space, are (by the addition of the axioms of groups I and II) exclusively the consequences of the six linear and plane axioms mentioned above. Hence, it is not necessary to assume the axiom of parallels in order to establish these facts.

If we take, in, addition to the axioms of congruence, the axiom of parallels, we can then easily establish the following propositions:

Theorem 19

If two parallel lines are cut by a third straight line, the alternate-interior angles and also the exterior-interior angles are congruent. Conversely, if the alternate-interior or the exterior-interior angles are congruent, the given lines are parallel.

Theorem 20

The sum of the angles of a triangle is two right angles.

Definitions

If M is an arbitrary point in the plane α, the totality of all points A, for which the segments MA are congruent to one another, is called *a circle*. M is called the *centre of the circle*.

From this definition can be easily deduced, with the help of the axioms of groups III and IV, the known properties of the circle; in particular, the possibility of constructing a circle through any three points not lying in a straight line, as also the congruence of all angles inscribed in the same segment of a circle, and the theorem relating to the angles of an inscribed quadrilateral.

§8. GROUP V: AXIOM OF CONTINUITY (ARCHIMEDEAN AXIOM)

This axiom makes possible the introduction into geometry of the idea of continuity. In order to state this axiom, we must first establish a convention concerning the equality of two segments. For this purpose, we can either base our idea of equality upon the axioms relating to the congruence of segments and define as "*equal*" the correspondingly congruent segments, or upon the basis of groups I and II, we may determine how, by suitable constructions (see Chap. V, § 24), a segment is to be laid off from a point of a given straight line so that a new, definite segment is obtained "*equal*" to it. In conformity with such a convention, the axiom of Archimedes may be stated as follows:

V.   *Let A_1 be any point upon a straight line between the arbitrarily chosen points A and B. Take the points A_2, A_3, A_4,.... so that A_1 lies between A and A_2, A_2 between A_1 and A_3, A_3 between A_2 and A_4 etc. Moreover, let the segments*

$$AA_1, A_1A_2, A_2A_3, A_3A_4, \ldots$$

be equal to one another. Then, among this series of points, there always exists a certain point A_n such that B lies between A and A_n.

The axiom of Archimedes is a linear axiom.

Remark³ To the preceeding five groups of axioms, we may add the following one, which, although not of a purely geometrical nature, merits particular attention from a theoretical point of view. It may be expressed in the following form:

Axiom of Completeness⁴ (*Vollstandigkeit*)

To a system of points, straight lines, and planes, it is impossible to add other elements in such a manner that the system thus generalized shall form a new geometry obeying all of the five groups of axioms. In other words, the elements of geometry form a system which is not susceptible of extension, if we regard the five groups of axioms as valid.

This axiom gives us nothing directly concerning the existence of limiting points, or of the idea of convergence. Nevertheless, it enables us to demonstrate Bolzano's theorem by virtue of which, for all sets of points situated upon a straight line between two definite points of the same line, there exists necessarily a point of condensation, that is to say, a limiting point. From a theoretical point of view, the value of this axiom is that it leads indirectly to the introduction of limiting points, and, hence, renders it possible to establish a one-to-one correspondence between the points of a segment and the system of real numbers. However, in what is to follow, no use will be made of the "axiom of completeness."

2

COMPATIBILITY AND MUTUAL
INDEPENDENCE OF THE AXIOMS

§9. COMPATIBILITY OF THE AXIOMS

The axioms, which we have discussed in the previous chapter and have divided into five groups, are not contradictory to one another; that is to say, it is not possible to deduce from these axioms, by any logical process of reasoning, a proposition which is contradictory to any of them. To demonstrate this, it is sufficient to construct a geometry where all of the five groups are fulfilled.

To this end, let us consider a domain Ω consisting of all of those algebraic numbers which may be obtained by beginning with the number one and applying to it a finite number of times the four arithmetical operations (addition, subtraction, multiplication, and division) and the operation $\sqrt{1 + \omega^2}$, where ω represents a number arising from the five operations already given.

Let us regard a pair of numbers (x, y) of the domain Ω as defining a point and the ratio of three such numbers $(u : v : w)$ of Ω, where u, v are not both equal to zero, as defining a straight line. Furthermore, let the existence of the equation

$$ux + vy + w = 0$$

express the condition that the point (x, y) lies on the straight line $(u : v : w)$. Then, as one readily sees, axioms I, 1–2 and III are fulfilled. The numbers of the domain Ω are all real numbers. If now we take into consideration the fact that these numbers may be arranged according to magnitude, we can easily make such necessary conventions concerning our points and straight lines as will also make the axioms of order (group II) hold. In fact, if (x_1, y_1), (x_2, y_2), (x_3, y_3), ... are any points whatever of a straight line, then this may be taken as their sequence on this straight line, providing the numbers $x_1, x_2, x_3, ...$, or the numbers $y_1, y_2, y_3, ...$, either all increase or decrease in the order of sequence given here. In order that axiom II, 5 shall be fulfilled, we have merely to assume that all points corresponding to values of x and y which make $ux + vy + w$ less than zero or greater than zero shall fall respectively upon the one side or upon the other side of the straight line $(u : v : w)$. We can easily convince ourselves that this convention is in accordance with those which precede, and by which the sequence of the points on a straight line has already been determined.

The laying off of segments and of angles follows by the known methods of analytical geometry. A transformation of the form

$$x' = x + a$$
$$y' = y + b$$

produces a translation of segments and of angles.

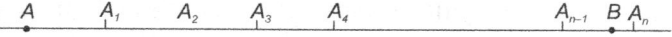

Figure 13

Furthermore, if, in the accompanying figure, we represent the point $(0, 0)$ by O and the point $(1, 0)$ by E, then, corresponding to a rotation of the angle *COE* about *O* as a center, any point (x, y) is transformed into another point (x', y') so related that

$$x' = \frac{a}{\sqrt{a^2 + b^2}} x - \frac{b}{\sqrt{a^2 + b^2}} y,$$

$$y' = \frac{b}{\sqrt{a^2 + b^2}} x + \frac{a}{\sqrt{a^2 + b^2}} y.$$

Since the number

$$\sqrt{a^2 + b^2} = a\sqrt{1 + \left(\frac{b}{a}\right)^2}$$

belongs to the domain Ω, it follows that, under the conventions which we have made, the axioms of congruence (group IV) are all fulfilled. The same is true of the axiom of Archimedes.

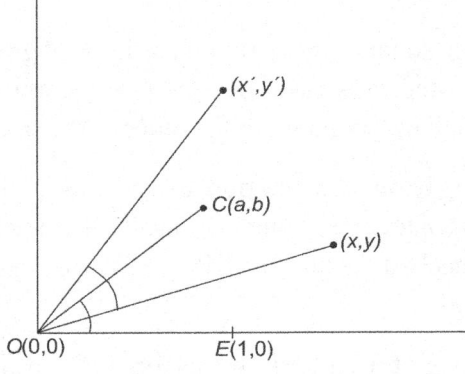

Figure 14

From these considerations, it follows that every contradiction resulting from our system of axioms must also appear in the arithmetic related to the domain Ω.

The corresponding considerations for the geometry of space present no difficulties.

If, in the preceding development, we had selected the domain of all real numbers instead of the domain Ω, we should have obtained likewise a geometry in which all of the axioms of groups I–V are valid. For the purposes of our demonstration, however, it was sufficient to take the domain Ω, containing on an enumerable set of elements.

§10. INDEPENDENCE OF THE AXIOMS OF PARALLELS (NON-EUCLIDEAN GEOMETRY)[5]

Having shown that the axioms of the above system are not contradictory to one another, it is of interest to investigate the question of their mutual independence. In fact, it may be shown that none of them can be deduced from the remaining ones by any logical process of reasoning.

First of all, so far as the particular axioms of groups I, II, and IV are concerned, it is easy to show that the axioms of these groups are each independent of the other of the same group.[6]

According to our presentation, the axioms of groups I and II form the basis of the remaining axioms. It is sufficient, therefore, to show that each of the groups II, IV, and V is independent of the others.

The first statement of the axiom of parallels can be demonstrated by aid of the axioms of groups I, II, and IV. In order to do this, join the given point A with any arbitrary point B of the straight line a. Let C be any other point of the given straight line. At the point A on AB, construct the angle ABC so that it shall lie in the same plane α as the point C, but upon the opposite side of *AB* from it. The straight line thus obtained through *A* does not meet the given straight line *a*; for, if it should cut it, say in the point *D*, and if we suppose *B* to be situated between *C* and *D*, we could then find on *a* a point *D′* so situated that *B* would lie between *D* and *D′*, and, moreover, so that we should have

$$AD \equiv BD'$$

Because of the congruence of the two triangles *ABD* and *BAD′*, we have also

$$\angle ABD \equiv \angle BAD',$$

and since the angles *ABD′* and *ABD* are supplementary, it follows from theorem 12 that the angles *BAD* and *BAD′* are also supplementary. This, however, cannot be true, as, by theorem 1, two straight lines cannot intersect in more than one point, which would be the case if *BAD* and *BAD′* were supplementary.

The second statement of the axiom of parallels is independent of all the other axioms. This may be most easily shown in the following well known manner. As the individual elements of a geometry of space, select the points, straight lines, and planes of the ordinary geometry as constructed in §9, and regard these elements as restricted in extent to the interior of a fixed sphere. Then, define the congruences of this geometry by aid of such linear transformations of the ordinary geometry as transform the fixed sphere into itself. By suitable conventions, we can make this *"non-euclidean geometry"* obey all of the axioms of our system except the axiom of Euclid (group III). Since the possibility of the ordinary geometry has already been established, that of the non-euclidean geometry is now an immediate consequence of the above considerations.

§11. INDEPENDENCE OF THE AXIOMS OF CONGRUENCE

We shall show the independence of the axioms of congruence by demonstrating that axiom IV, 6, or what amounts to the same thing, that the first theorem of congruence for triangles (theorem 10) cannot be deduced from the remaining axioms I, II, III, IV 1–5, V by any logical process of reasoning.

Select, as the points, straight lines, and planes of our new geometry of space, the points, straight lines, and planes of ordinary geometry, and define the laying off of an angle as in ordinary geometry, for example, as explained in §9. We will, however, define the laying off of segments in another manner. Let A_1, A_2 be two points which, in ordinary geometry, have the co-ordinates x_1, y_1, z_1 and x_2, y_2, z_2, respectively. We will now define the length of the segment $A_1 A_2$ as the positive value of the expression

$$\sqrt{(x_1 - x_2 + y_1 - y_2)^2 + (y_1 - y_2)^2 + (z_1 - z_2)^2}$$

and call the two segments $A_1 A_2$ and $A_1' A_2'$ congruent when they have equal lengths in the sense just defined.

It is at once evident that, in the geometry of space thus defined, the axioms I, II, III, IV 1–2, 4–5, V are all fulfilled.

In order to show that axiom IV, 3 also holds, we select an arbitrary straight line a and upon it three points A_1, A_2, A_3 so that A_2 shall lie between A_1 and A_3. Let the points x, y, z of the straight line a be given by means of the equations

$$x = \lambda t + \lambda',$$
$$y = \mu t + \mu',$$
$$z = \nu t + \nu',$$

where $\lambda, \lambda', \mu, \mu', \nu, \nu'$ represent certain constants and T is a parameter. If $t_1, t_2 \ (<t_1), t_3 \ (<t_2)$ are the values of the parameter corresponding to the points A_1, A_2, A_3 we have as the lengths of the three segments $A_1A_2 A_2A_3$ and A_1A_3 respectively, the following values:

$$(t_1 - t_2)\left|\sqrt{(\lambda + \mu)^2 + \mu^2 + \nu^2}\right|$$

$$(t_2 - t_3)\left|\sqrt{(\lambda + \mu)^2 + \mu^2 + \nu^2}\right|$$

$$(t_1 - t_3)\left|\sqrt{(\lambda + \mu)^2 + \mu^2 + \nu^2}\right|$$

Consequently, the length of A_1A_3 is equal to the sum of the lengths of the segments A_1A_2 and A_2A_3. But this result is equivalent to the existence of axiom IV, 3.

Axiom IV, 6, or rather the first theorem of congruence for triangles, is not always fulfilled in this geometry. Consider, for example, in the plane $z = 0$, the four points

O, having the co-ordinates $x = 0, y = 0$

A, having the co-ordinates $x = 1, y = 0$

B, having the co-ordinates $x = 0, y = 1$

C, having the co-ordinates $x = \dfrac{1}{2}, y = \dfrac{1}{2}$

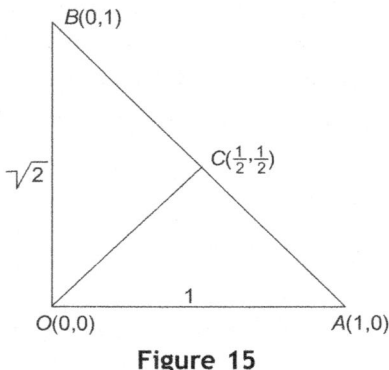

Figure 15

Then, in the right triangles OAC and OBC, the angles at C as also the adjacent sides AC and BC are respectively congruent; for, the side OC is common to the two triangles and the sides AC and BC have the same length, namely, $\frac{1}{2}$. However, the third sides OA and OB have the lengths 1 and $\sqrt{2}$, respectively, and are not, therefore, congruent. It is not difficult to find in this geometry two triangles for which axiom IV, 6, itself is not valid.

§12. INDEPENDENCE OF THE AXIOM OF CONTINUITY (NON-ARCHIMEDEAN GEOMETRY)

In order to demonstrate the independence of the axiom of Archimedes, we must produce a geometry in which all of the axioms are fulfilled with the exception of the one in question.[7]

For this purpose, we construct a domain $\Omega(t)$ of all those algebraic functions of t which may be obtained from t by means of the four arithmetical operations of addition, subtraction, multiplication, division, and the fifth operation $\sqrt{1 + \omega^2}$, where ω represents any function arising from the application of these five operations. The elements of $\Omega(t)$—just as was previously the case for Ω—constitute an enumerable set. These five operations may all be performed without introducing imaginaries, and that in only one way. The domain $\Omega(t)$ contains, therefore, only real, single-valued functions of t.

Let c be any function of the domain $\Omega(t)$. Since this function c is an algebraic function of t, it can in no case vanish for more than a finite number of values of t, and, hence, for sufficiently large positive values of t, it must remain always positive or always negative.

Let us now regard the functions of the domain $\Omega(t)$ as a kind of complex numbers. In the system of complex numbers thus defined, all of the ordinary rules of operation evidently hold. Moreover, if $a,\ b$ are any two distinct numbers of this system, then a is said to be greater than, or less than, b (written $a > b$ or $a < b$) according as the difference $c = a - b$ is always positive or always negative for sufficiently large values of t. By the adoption of this convention for the numbers of our system, it is possible to arrange them according to their magnitude in a manner analogous to that employed for real numbers. We readily see also that, for this system of complex numbers, the validity of an inequality is not destroyed by adding the same or equal numbers to both members, or by multiplying both members by the same number, providing it is greater than zero.

If n is any arbitrary positive integral rational number, then, for the two numbers n and t of the domain $\Omega(t)$, the inequality $n < t$ certainly holds; for, the difference $n - t$, considered as a function of t, is always negative for sufficiently large values of t. We express this fact in the following manner: The two numbers l and t of the domain $\Omega(t)$, each of which is greater than zero, possess the property that any multiple whatever of the first always remains smaller than the second.

From the complex numbers of the domain $\Omega(t)$, we now proceed to construct a geometry in exactly the same manner as in §9, where we took as the basis of our consideration the algebraic numbers of the domain Ω. We will regard a system of three numbers (x, y, z) of the domain $\Omega(t)$ as defining a point, and the ratio of any four such numbers $(u : v : w : r)$, where u, v, w are not all zero, as defining a plane. Finally, the existence of the equation

$$xu + yv + zw + r = 0$$

shall express the condition that the point (x, y, z) lies in the plane $(u : v : w : r)$. Let the straight line be defined in our geometry as the totality of all the points lying simultaneously in the same two planes. If now we adopt conventions corresponding to those of §9 concerning the arrangement of elements and the laying off of angles and of segments, we shall obtain a *"non-archimedean"* *geometry* where, as the properties of the complex number system already investigated show, all of the axioms, with the exception of that of Archimedes, are fulfilled. In fact, we can lay off successively the segment 1 upon the segment t an arbitrary number of times without reaching the end point of the segment t, which is a contradiction to the axiom of Archimedes.

3

THE THEORY OF PROPORTION[8]

§13. COMPLEX NUMBER-SYSTEMS

At the beginning of this chapter, we shall present briefly certain preliminary ideas concerning complex number systems which will later be of service to us in our discussion.

The real numbers form, in their totality, a system of things having the following properties:

Theorems of Connection (1–12)

1. From the number a and the number b, there is obtained by "addition" a definite number c, which we express by writing
    $$a + b = c \text{ or } c = a + b.$$

2. There exists a definite number, which we call 0, such that, for every number a, we have
    $$a + 0 = a \text{ and } 0 + a = a.$$

3. If a and b are two given numbers, there exists one and only one number x, and also one and only one number y, such that we have respectively
    $$a + x = b, \ y + a = b.$$

4. From the number a and the number b, there may be obtained in another way, namely, by "multiplication," a definite number c, which we express by writing

$$ab = c \text{ or } c = ab.$$

5. There exists a definite number, called 1, such that, for every number a, we have

$$a.1 = a \text{ and } 1.a = a.$$

6. If a and b are any arbitrarily given numbers, where a is different from 0, then there exists one and only one number x and also one and only one number y such that we have respectively

$$ax = b, \ ya = b.$$

If a, b, c are arbitrary numbers, the following laws of operation always hold:

7. $a +(b + c) = (a + b) + c$

8. $a + b = b + a$

9. $a(bc) = (ab)c$

10. $a(b + c) = ab + ac$

11. $(a + b)c = ac + bc$

12. $ab = ba$

Theorems of Order (13–16)

13. If a, b are any two distinct numbers, one of these, say a, is always greater (>) than the other. The other number is said to be the smaller of the two. We express this relation by writing $a > b$ and $b < a$.

14. If $a > b$ and $b > c$, then is also $a > c$.

15. If $a > b$, then is also $a + c > b + c$ and $c + a > c + b$.

16. If $a > b$ and $c > 0$, then is also $ac > bc$ and $ca > cb$.

Theorem of Archimedes (17)

17. If a, b are any two arbitrary numbers, such that $a > 0$ and $b > 0$, it is always possible to add a to itself a sufficient number of times so that the resulting sum shall have the property that

$$a + a + a + \cdots + a > b.$$

A system of things possessing only a portion of the above properties (1–17) is called a *complex number system*, or simply a *number system*. A number system is called *archimedean*, or *non-archimedean*, according as it does, or does not, satisfy condition (17).

Not every one of the properties (1–17) given above is independent of the others. The problem arises to investigate the logical dependence of these properties. Because of their great importance in geometry, we shall, in §§32, 33, pp. 65–68, answer two definite questions of this character. We will here merely call attention to the fact that, in any case, the last of these conditions (17) is not a consequence of the remaining properties, since, for example, the complex number system $\Omega(t)$, considered in §12, possesses all of the properties (1–16), but does not fulfil the law stated in (17).

§14. DEMONSTRATION OF PASCAL'S THEOREM

In this and the following chapter, we shall take as the basis of our discussion all of the plane axioms with the exception of the axiom of Archimedes; that is to say, the axioms I, 1–2 and II–IV. In the present chapter, we propose, by aid of these axioms, to establish Euclid's theory of proportion; that is, *we shall establish it for the plane and that independently of the axiom of Archimedes*.

For this purpose, we shall first demonstrate a proposition which is a special case of the well known theorem of Pascal usually considered in the theory of conic sections, and which we shall hereafter, for the sake of brevity, refer to simply as Pascal's theorem. This theorem may be stated as follows:

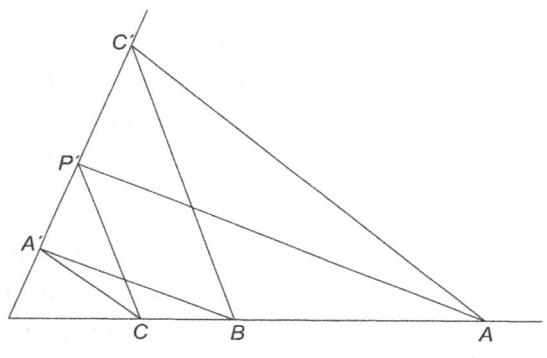

Figure 16

Theorem 21 (Pascal's Theorem)

Given the two sets of points A, B, C and A', B', C' so situated respectively upon two intersecting straight lines that none of them fall at the intersection of these lines. If CB' is parallel to BC' and CA' is also parallel to AC', then BA' is parallel to AB'.[9]

In order to demonstrate this theorem, we shall first introduce the following notation. In a right triangle, the base a is uniquely determined by the hypotenuse c and the base angle α included by a and c. We will express this fact briefly by writing

$$a = \alpha c.$$

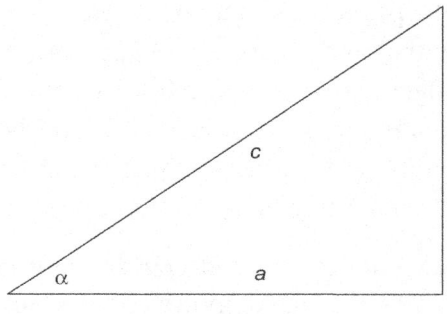

Figure 17

Hence, the symbol αc always represents a definite segment, providing c is any given segment whatever and α is any given acute angle.

Furthermore, if c is any arbitrary segment and α, β are any two acute angles whatever, then the two segments $\alpha\beta c$ and $\beta\alpha c$ are always congruent; that is, we have

$$\alpha\beta c = \beta\alpha c,$$

and, consequently, the symbols α and β are interchangeable.

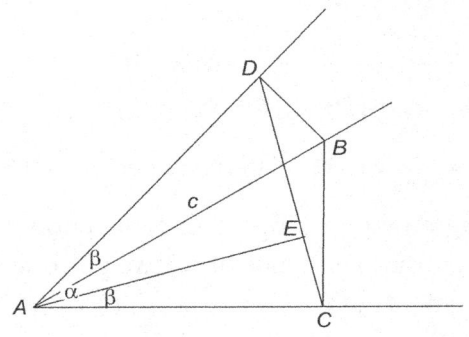

Figure 18

In order to prove this statement, we take the segment $c = AB$, and with A as a vertex lay off upon the one side of this segment the angle α and upon the other the angle β. Then, from the point B, let fall upon the opposite sides of the α and β the perpendiculars BC and BD, respectively. Finally, join C with D and let fall from A the perpendicular AE upon CD.

Since the two angles ACB and ADB are right angles, the four points A, B, C, D are situated upon a circle. Consequently, the angles ACD and ABD, being inscribed in the same segment of the circle, are congruent. But the angles ACD and CAE, taken together, make a right angle, and the same is true of the two angles ABD and BAD. Hence, the two angles CAE and BAD are also congruent; that is to say,

$$\angle CAE \equiv \beta$$

and, therefore,

$$\angle DAE \equiv \alpha$$

From these considerations, we have immediately the following congruences of segments:

$$\beta c \equiv AD, \qquad \alpha c \equiv AC,$$
$$\alpha\beta c \equiv \alpha(AD) \equiv AE, \quad \beta\alpha c \equiv \beta(AC) \equiv AE$$

From these, the validity of the congruence in question follows.

Returning now to the figure in connection with Pascal's theorem, denote the intersection of the two given straight lines by O and the segments OA, OB, OC, OA', OB', OC', CB', BC', CA', AC', BA', AB' by $a, b, c, a', b', c', l, l^*, m, m^*, n, n^*$, respectively.

Let fall from the point O a perpendicular upon each of the segments l, m, n. The perpendicular to l will form with the straight lines OA and OA' acute angles, which we shall denote by λ' and λ, respectively. Likewise, the perpendiculars to m and n form with these same lines OA and OA' acute angles, which we shall denote by μ', μ and ν', ν, respectively. If we now express, as indicated above, each of these perpendiculars in terms of the hypotenuse and base angle, we have the three following congruences of segments:

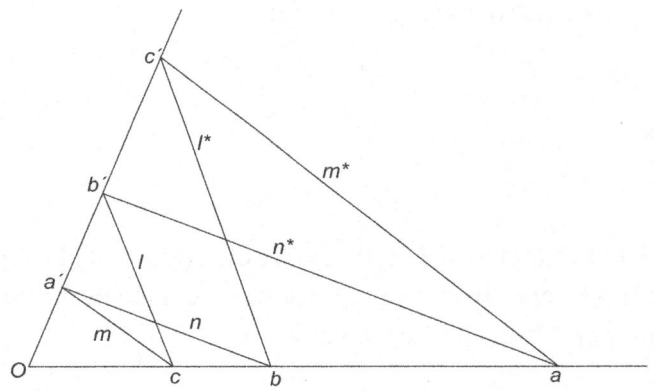

Figure 19

(1) $\lambda b' \equiv \lambda' c$

(2) $\mu a' \equiv \mu' c$

(3) $\nu a' \equiv \nu' b$

But since, according to our hypothesis, l is parallel to l^* and m is parallel to m^*, the perpendiculars from O falling upon l^* and m^* must coincide with the perpendiculars from the same point falling upon l and m, and consequently, we have

(4) $\lambda c' \equiv \lambda' b$,

(5) $\mu c' \equiv \mu' a$.

Multiplying both members of congruence (3) by the symbol $\lambda' \mu$, and remembering that, as we have already seen, the symbols in question are commutative, we have

$$v\lambda'\mu a' \equiv v'\mu\lambda' b.$$

In this congruence, we may replace $\mu a'$ in the first member by its value given in (2) and $\lambda' b$ in the second member by its value given in (4), thus obtaining as a result

$$v\lambda'\mu' c \equiv v'\mu\lambda c',$$

or

$$v\mu'\lambda' c \equiv v'\lambda\mu c'.$$

Here again in this congruence we can, by aid of (1), replace $\lambda' c$ by $\lambda b'$, and, by aid of (5), we may replace in the second member $\mu c'$ by $\mu' a$. We then have

$$v\mu'\lambda b' \equiv v'\lambda\mu' a,$$

or

$$\lambda\mu' v b' \equiv \lambda\mu' v' a.$$

Because of the significance of our symbols, we can conclude at once from this congruence that

$$\mu' v b' \equiv \mu' v' a,$$

and, consequently, that

(6) $v b' \equiv v' a$

If now we consider the perpendicular let fall from O upon n and draw perpendiculars to this same line from the points A and B', then congruence (6) shows that the feet of the last two perpendiculars must coincide; that is to say, the straight line $n^* = AB'$ makes a right angle with the perpendicular to n and, consequently, is parallel to n. This establishes the truth of Pascal's theorem.

Having given any straight line whatever, together with an arbitrary angle and a point lying outside of the given line, we can, by constructing the given angle and drawing a parallel line, find a straight line passing through the given point and cutting the given straight line at the given angle. By means of this construction, we can demonstrate Pascal's theorem in the following very simple manner, for which, however, I am indebted to another source.

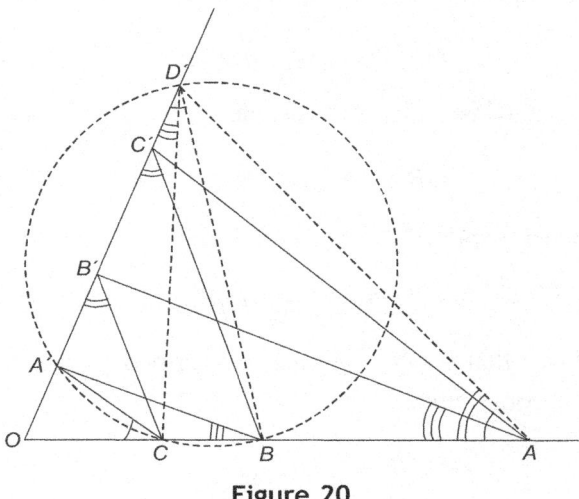

Figure 20

Through the point B, draw a straight line cutting OA' in the point D' and making with it the angle OCA', so that the congruence

$$(1^*) \qquad \angle OCA' \equiv \angle OD'B$$

is fulfilled. Now, according to a well known property of circles, $CBD'A'$ is an inscribed quadrilateral, and, consequently, by aid of the theorem concerning the congruence of angles inscribed in the same segment of a circle, we have the congruence

$$(2^*) \qquad \angle OBA' \equiv \angle OD'C.$$

Since, by hypothesis, CA' and AC' are parallel, we have

$$(3^*) \qquad \angle OCA' \equiv \angle OAC',$$

and from (1^*) and (3^*) we obtain the congruence

$$\angle OD'B \equiv \angle OAC'.$$

However, $BAD'C'$ is also an inscribed quadrilateral, and, consequently, by virtue of the theorem relating to the angles of such a quadrilateral, we have the congruence

$$(4^*) \qquad \angle OAD' \equiv \angle OC'B.$$

But as CB' is, by hypothesis, parallel to BC', we have also

$$(5^*) \qquad \angle OB'C \equiv \angle OC'B.$$

From (4^*) and (5^*), we obtain the congruence

$$\angle OAD' \equiv \angle OB'C,$$

which shows that $CAD'B'$ is also an inscribed quadrilateral, and, hence, the congruence

$$(6^*) \qquad \angle OAB' \equiv \angle OD'C,$$

is valid. From (2^*) and (6^*), it follows that

$$\angle OBA' \equiv \angle OAB',$$

and this congruence shows that BA' and AB' are parallel as Pascal's theorem demands. In case D' coincides with one of the points A', B', C', it is necessary to make a modification of this method, which evidently is not difficult to do.

§15. AN ALGEBRA OF SEGMENTS, BASED UPON PASCAL'S THEOREM

Pascal's theorem, which was demonstrated in the last section, puts us in a position to introduce into geometry a method of calculating with segments, in which all of the rules for calculating with real numbers remain valid without any modification.

Instead of the word "congruent" and the sign $\equiv$, we make use, in the algebra of segments, of the word "equal" and the sign $=$.

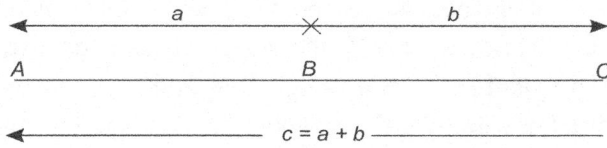

Figure 21

If A, B, C are three points of a straight line and if B lies between A and C, then we say that $c = AC$ is the *sum* of the two segments $a = AB$ and $b = BC$. We indicate this by writing

$$c = a + b.$$

The segments a and b are said to be smaller than c, which fact we indicate by writing

$$a < c,\ b < c.$$

On the other hand, c is said to be larger than a and b, and we indicate this by writing

$$c > a,\ c > b.$$

From the linear axioms of congruence (axioms IV, 1–3), we easily see that, for the above definition of addition of segments, the associative law

$$a + (b + c) = (a + b) + c,$$

as well as the commutative law

$$a + b = b + a$$

is valid.

In order to define geometrically the product of two segments a and b, we shall make use of the following construction. Select any convenient segment, which, having been selected, shall remain constant throughout the discussion, and denote the same by 1. Upon the one side of a right angle, lay off from the vertex O the segment 1 and also the segment b. Then, from O lay off upon the other side of the right angle the segment a. Join the extremities of the segments 1 and a by a straight line, and from the extremity of b draw a line parallel to this straight line. This parallel will cut off from the other side of the right angle a segment c. We call this segment c the *product* of the segments a and b, and indicate this relation by writing

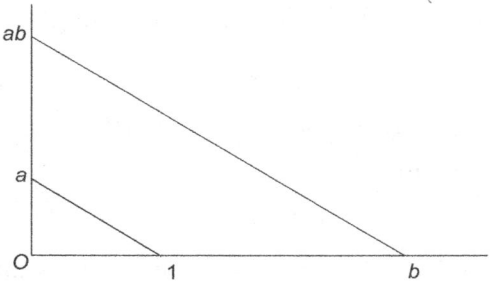

Figure 22

$$c = ab.$$

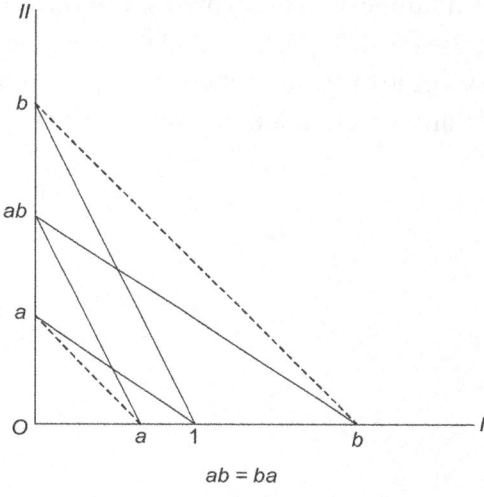

ab = ba

Figure 23

We shall now demonstrate that, for this definition of the multiplication of segments, the *commutative* law

$$ab = ba$$

holds. For this purpose, we construct in the above manner the product *ab*. Furthermore, lay off from 0 upon the first side (I) of the right angle the segment *a* and upon the other side (II) the segment *b*. Connect by a straight line the extremity of the segment 1 with the extremity of *b*, situated on II, and draw through the endpoint of *a*, on I, a line parallel to this straight line. This parallel will determine, by its intersection with the side II, the segment *ba*. But, because the two dotted lines are, by Pascal's theorem, parallel, the segment *ba* just found coincides with the segment *ab* previously constructed, and our proposition is established. In order to show that the *associative* law

$$a(bc) = (ab)c$$

holds for the multiplication of segments, we construct first of all the segment $d = be$, then da, after that the segment $e = ba$, and finally ec. By virtue of Pascal's theorem, the extremities of the segments da and ec coincide, as may be clearly seen from figure 24.

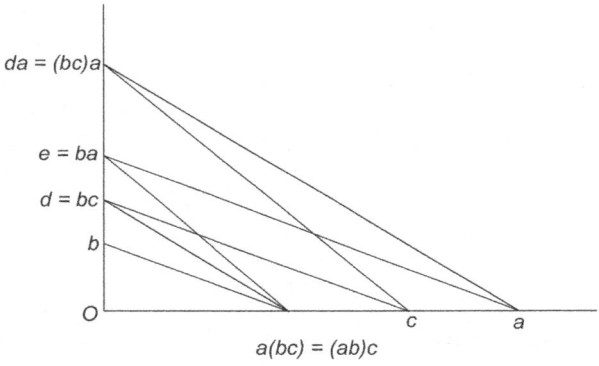

Figure 24

If now we apply the commutative law which we have just demonstrated, we obtain the above formula, which expresses the associative law for the multiplication of two segments.

Finally, *the distributive law*

$$a(b + c) = ab + ac$$

also holds for our algebra of segments. In order to demonstrate this, we construct the segments, *ab, ac*, and $a(b + c)$, and draw through the extremity of the segment c (Figure 25) a straight line parallel to the other side of the right angle. From the congruence of the two right-angled triangles which are shaded in the figure and the application of the theorem relating to the equality of the opposite sides of a parallelogram, the desired result follows.

If b and c are any two arbitrary segments, there is always a segment a to be found such that $c = ab$. This segment a is denoted by $\dfrac{c}{b}$ and is called the *quotient* of c by b.

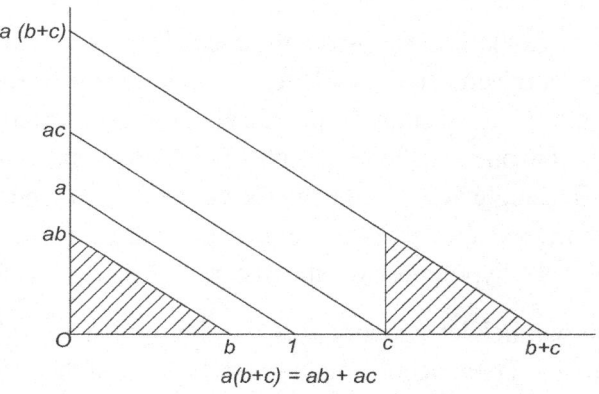

$$a(b+c) = ab + ac$$

Figure 25

§16. PROPORTION AND THE THEOREMS OF SIMILITUDE

By aid of the preceding algebra of segments, we can establish Euclid's theory of proportion in a manner free from objections and without making use of the axiom of Archimedes.

If a, b, a', b' are any four segments whatever, the proportion

$$a : b = a' : b'$$

expresses nothing else than the validity of equation

$$ab' = ba'$$

Definition

Two triangles are called *similar* when the corresponding angles are congruent.

Theorem 22

If a, b and a', b' are homologous sides of two similar triangles, we have the proportion

$$a : b = a' : b'$$

Proof   We shall first consider the special case where the angle included between a and b and the one included between a' and b' are right angles. Moreover, we shall assume that the two triangles are laid off in one and the same right angle. Upon one of the sides of this right angle, we lay off from the vertex 0 the segment 1, and through the extremity of this segment, we draw a straight line parallel to the hypotenuses of the two triangles.

This parallel determines upon the other side of the right angle a segment e. Then, according to our definition of the product of two segments, we have

$$b = ea, \ b' = ea',$$

from which we obtain

$$ab' = ba',$$

that is to say,

$$a : b = a' : b'.$$

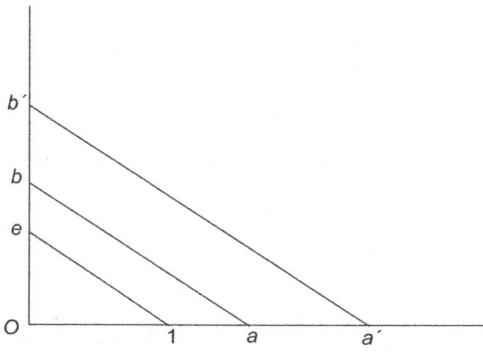

Figure 26

Let us now return to the general case. In each of the two similar triangles, find the point of intersection of the bisectors of the three angles. Denote these points by S and S'. From these

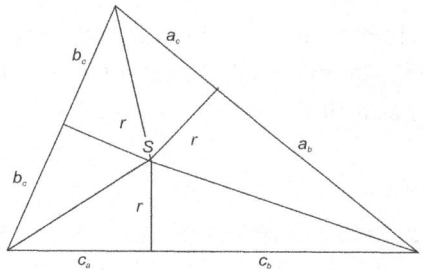

Figure 27

points let fall upon the sides of the triangles the perpendiculars r and r', respectively. Denote the segments thus determined upon the sides of the triangles by

$$a_b, a_c, b_c, b_a, c_a, c_b$$

and

$$a'_b, a'_c, b'_c, b'_a, c'_a, c'_b,$$

respectively. The special case of our proposition, demonstrated above, gives us then the following proportions:

$$a_b : r = a'_b : r', \qquad b_c : r = b'_c : r',$$
$$a_c : r = a'_c : r', \qquad b_a : r = b'_a : r'.$$

By aid of the distributive law, we obtain from these proportions the following:

$$a : r = a' : r, \quad b : r = b' : r'.$$

Consequently, by virtue of the commutative law of multiplication, we have

$$a : b = a' : b'.$$

From the theorem just demonstrated, we can easily deduce the fundamental theorem in the theory of proportion. This theorem may be stated as follows:

Theorem 23

If two parallel lines cut from the sides of an arbitrary angle the segments a, b and a', b' respectively, then we have always the proportion

$$a : b = a' : b'.$$

Conversely, if the four segments a, b, a', b fulfill this proportion and if a, a' and b, b' are laid off upon the two sides respectively of an arbitrary angle, then the straight lines joining the extremities of a and b and of a' and b' are parallel to each other.

§17. EQUATIONS OF STRAIGHT LINES AND OF PLANES

To the system of segments already discussed, let us now add a second system. We will distinguish the segments of the new system from those of the former one by means of a special sign, and will call them "*negative*" segments in contradistinction to the "*positive*" segments already considered. If we introduce also the segment O, which is determined by a single point, and make other appropriate conventions, then all of the rules deduced in §13 for calculating with real numbers will hold equally well here for calculating with segments. We call special attention to the following particular propositions:

We have always $a \cdot 1 = 1 \cdot a = a$.

If $a.b = 0$, then either $a = 0$, or $b = 0$.

If $a > b$ and $c > 0$, then $ac > bc$.

In a plane α, we now take two straight lines cutting each other in O at right angles as the fixed axes of rectangular co-ordinates, and lay off from O upon these two straight lines the arbitrary segments x and y. We lay off these segments upon the one side or upon the other side of O, according as they are positive or negative. At the extremities of x and y, erect perpendiculars and determine the point P of their intersection. The segments x and y are called the co-ordinates of P. Every point of the plane α is uniquely determined by its co-ordinates x, y, which may be positive, negative, or zero.

Let l be a straight line in the plane α, such that it shall pass through O and also through a point C having the co-ordinates a, b. If x, y are the co-ordinates

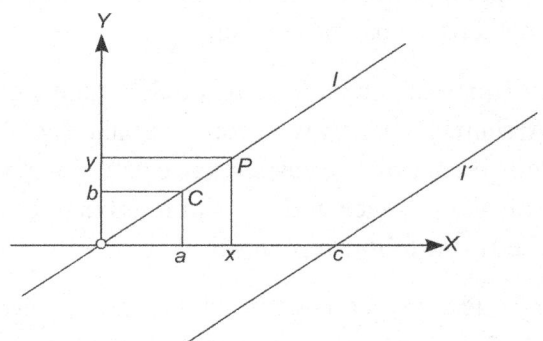

Figure 28

of any point on l, it follows at once from theorem 22 that

$$a : b = x : y,$$

or

$$bx - ay = 0,$$

is the equation of the straight line l. If l' is a straight line parallel to l and cutting off from the x-axis the segment c, then we may obtain the equation of the straight line l' by replacing, in the equation for l, the segment x by the segment $x - c$. The desired equation will then be of the form

$$bx - ay - bc = 0.$$

From these considerations, we may easily conclude, independently of the axiom of Archimedes, that every straight line of a plane is represented by an equation which is linear in the co-ordinates x, y, and, conversely, every such linear equation represents a straight line when the co-ordinates are segments appertaining to the geometry in question. The corresponding results for the geometry of space may be easily deduced.

The remaining parts of geometry may now be developed by the usual methods of analytic geometry.

So far in this chapter, we have made absolutely no use of the axiom of Archimedes. If now we assume the validity of this axiom, we can arrange a definite correspondence between the points on any straight line in space and the real numbers. This may be accomplished in the following manner.

We first select on a straight line any two points, and assign to these points the numbers 0 and 1. Then, bisect the segment $(0, 1)$ thus determined and denote the middle point by the number $\frac{1}{2}$. In the same way, we denote the middle of $\left(0, \frac{1}{2}\right)$ by $\frac{1}{4}$, etc. After applying this process n times, we obtain a point which corresponds to $\frac{1}{2^n}$. Now, lay off m times in both directions from the point O the segment $\left(O, \frac{1}{2^n}\right)$. We obtain in this manner a point corresponding to the numbers $\frac{m}{2^n}$ and $-\frac{m}{2^n}$. From the axiom of Archimedes, we now easily see that, upon the basis of this association, to each arbitrary point of a straight line there corresponds a single, definite, real number, and, indeed, such that this correspondence possesses the following property: If A, B, C are any three points on a straight line and α, β, γ are the corresponding real numbers, and, if B lies between A and C, then one of the inequalities,

$$\alpha < \beta < \gamma \text{ or } \alpha > \beta > \gamma,$$

is always fulfilled.

From the development given in §9, p. 15, it is evident, that to every number belonging to the field of algebraic numbers Ω, there must exist a corresponding point upon the straight line. Whether to every real number there corresponds a point cannot in general be established, but depends upon the geometry to which we have reference.

However, it is always possible to generalize the original system of points, straight lines, and planes by the addition of "ideal" or "irrational" elements, so that, upon any straight line of the corresponding geometry, a point corresponds without exception to every system of three real numbers. By the adoption of suitable conventions, it may also be seen that, in this generalized geometry, all of the axioms I–V are valid. This geometry, generalized by the addition of irrational elements, is nothing else than the ordinary analytic geometry of space.

4

THE THEORY OF PLANE AREAS

§18. EQUAL AREA AND EQUAL CONTENT OF POLYGONS.[10]

We shall base the investigations of the present chapter upon the same axioms as were made use of in the last chapter, §§ 13–17, namely, upon the plane axioms of all the groups, with the single exception of the axiom of Archimedes. This involves then the axioms I, 1–2 and II–IV.

The theory of proportion as developed in §§ 13–17 together with the algebra of segments introduced in the same chapter, puts us now in a position to establish Euclid's theory of areas by means of the axioms already mentioned; that is to say, *for the plane geometry, and that independently of the axiom of Archimedes.*

Since, by the development given in the last chapter, pp. 23–35, the theory of proportion was made to depend essentially upon Pascal's theorem (theorem 21), the same may then be said here of the theory of areas. This manner of establishing the theory of areas seems to me a very remarkable application of Pascal's theorem to elementary geometry.

If we join two points of a polygon P by any arbitrary broken line lying entirely within the polygon, we shall obtain two new polygons P_1 and P_2 whose interior points all lie within P. We say that P is *decomposed into* P_1 and P_2, or that the polygon P is *composed of* P_1 and P_2.

Definition

Two polygons are said to be of *equal area* when they can be decomposed into a finite number of triangles which are respectively congruent to one another in pairs.

Definition

Two polygons are said to be of *equal content* when it is possible, by the addition of other polygons having equal area, to obtain two resulting polygons having equal area.

From these definitions, it follows at once that by combining polygons having equal area, we obtain as a result polygons having equal area. However, if from polygons having equal area we take polygons having equal area, we obtain polygons which are of equal content.

Furthermore, we have the following propositions:

Theorem 24

If each of two polygons P_1 and P_2 is of equal area to a third polygon P_3 then P_1 and P_2 are themselves of equal area. If each of two polygons is of equal content to a third, then they are themselves of equal content.

Proof   By hypothesis, we can so decompose each of the polygons P_1 and P_2 into such a system of triangles that any triangle of either of these systems will be congruent to the corresponding triangle of a system into which P_3 may be decomposed. If we consider simultaneously the two decompositions of P_3 we see that, in general, each triangle of the one decomposition is broken up into polygons by the segments which belong to the other decomposition.

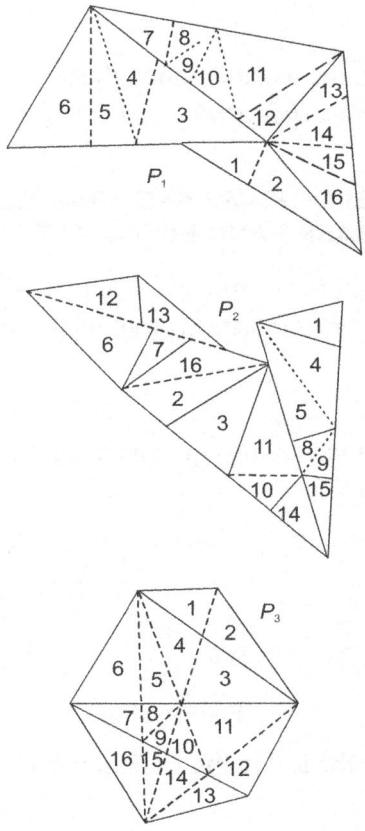

Figure 29

Let ms add to these segments a sufficient number of others to reduce each of these polygons to triangles, and apply the two resulting methods of decompositions to P_1 and P_2, thus breaking them up into corresponding triangles. Then, evidently the two polygons P_1 and P_2 are each decomposed into the same number of triangles, which are respectively congruent by pairs. Consequently, the two polygons are, by definition, of equal area.

The proof of the second part of the theorem follows without difficulty.

We define, in the usual manner, the terms: *rectangle, base* and *height of a parallelogram, base* and *height of a triangle.*

§19. PARALLELOGRAMS AND TRIANGLES HAVING EQUAL BASES AND EQUAL ALTITUDES

The well known reasoning of Euclid, illustrated by the accompanying figure, furnishes a proof for the following theorem:

Theorem 25

Two parallelograms having equal bases and equal altitudes are also of equal content.

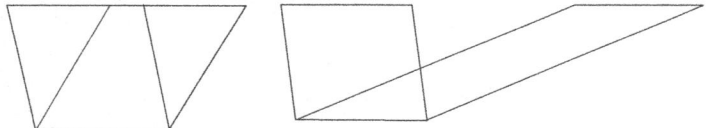

Figure 30

We have also the following well known proposition:

Theorem 26

Any triangle *ABC* is always of equal area to a certain parallelogram having an equal base and an altitude half as great as that of the triangle.

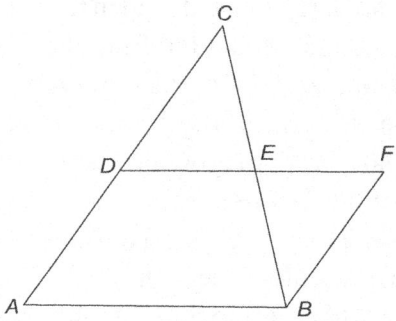

Figure 31

Proof Bisect AC in D and BC in E, and extend the line DE to F, making EF equal to DE. Then, the triangles DEC and FBE are congruent to each other, and, consequently, the triangle ABC and the parallelogram $ABFD$ are of equal area.

From theorems 25 and 26, we have at once, by aid of theorem 24, the following proposition.

Theorem 27

Two triangles having equal bases and equal altitudes have also equal content.

It is usual to show that two triangles having equal bases and equal altitudes are always of equal area. It is to be remarked, however, that *this demonstration cannot be made without the aid of the axiom of Archimedes.* In fact, we may easily construct in our non-archimedean geometry (see § 12, p. 21) two triangles so that they shall have equal bases and equal altitudes and, consequently, by theorem 27, must be of equal content, but which are not, however, of equal area. As such an example, we may take the two triangles ABC and ABD having each the base $AB = 1$ and the altitude 1, where the vertex of the first triangle is situated perpendicularly above *A*, and in the second triangle the foot *F* of the perpendicular let fall from the vertex *D* upon the base is so situated that *AF* = *t*. The remaining propositions of elementary geometry concerning the equal content of polygons, and in particular the pythagorean theorem, are all simple consequences of the theorems which we have already given. In the further development of the theory of area, we meet, however, with an essential difficulty. In fact, the discussion so far leaves it still in doubt whether all polygons are not, perhaps, of equal content. In this case, all of the propositions which we have given would be devoid of meaning and hence of no value. Furthermore, the more general question also arises as to whether two rectangles of equal content and having one side in common, do not also have their other sides congruent; that is to say, whether a rectangle is not definitely determined by means of a side and its area. As a closer investigation shows, in order to answer this question, we need to make use of the converse of theorem 27. This may be stated as follows:

Theorem 28

If two triangles have equal content and equal bases, they have also equal altitudes.

This fundamental theorem is to be found in the first book of Euclid's Elements as proposition 39. In the demonstration of this theorem, however, Euclid appeals to the general proposition relating to magnitudes: *"Καὶ τὸ ὅλου τοῦ μέρους μεῖζόν ἐστιν"*—a method of procedure which amounts to the same thing as introducing a new geometrical axiom concerning areas.

It is now possible to establish the above theorem and hence the theory of areas in the manner we have proposed, that is to say, with the help of the plane axioms and without making use of the axiom of Archimedes. In order to show this, it is necessary to introduce the idea of the measure of area.

§20. THE MEASURE OF AREA OF TRIANGLES AND POLYGONS

Definition

If in a triangle ABC, having the sides a, b, c, we construct the two altitudes $h_a = AD$, $h_b = BE$, then, according to theorem 22, it follows from the similarity of the triangles BCE and ACD, that we have the proportion

$$a : h_b = b : h_a;$$

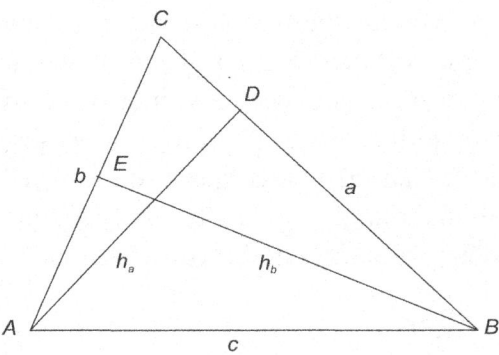

Figure 32

that is, we have

$$a \cdot h_a = b \cdot h_b$$

This shows that the product of the base and the corresponding altitude of a triangle is the same whichever side is selected as the base. The half of this product of the base and the altitude of a triangle Δ is called the *measure of area of the triangle* Δ and we denote it by $F(\Delta)$. A segment joining a vertex of a triangle with a point of the opposite side is called *a transversal*. A transversal divides the given triangle into two others having the same altitude and having bases which lie in the same straight line. Such a decomposition is called a *transversal decomposition of the triangle.*

Theorem 29

If a triangle Δ is decomposed by means of arbitrary straight lines into a finite number of triangles Δ_k, then the measure of area of Δ is equal to the sum of the measures of area of the separate triangles Δ_k.

Proof   From the distributive law of our algebra of segments, it follows immediately that the measure of area of an arbitrary triangle is equal to the sum of the measures of area of two such triangles as arise from any transversal decomposition of the given triangle. The repeated application of this proposition shows that the measure of area of any triangle is equal to the sum of the measures of area of all the triangles arising by applying the transversal decomposition an arbitrary number of times in succession.

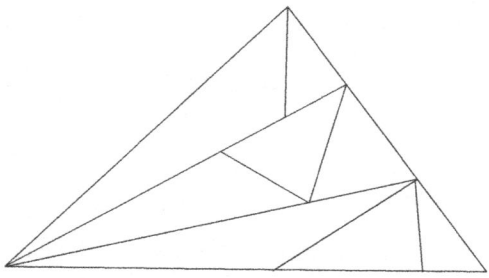

Figure 33

In order to establish the corresponding proof for an arbitrary decomposition of the triangle Δ into the triangles Δ_k, draw from the vertex A of the given triangle Δ a transversal through each of the points of division of the required decomposition; that is to say, draw a transversal through each vertex of the triangles Δ_k. By means of these transversals, the given triangle Δ is decomposed into certain triangles Δ_t. Each of these triangles Δ_t is broken up by the segments which determined this decomposition into certain triangles and quadrilaterals. If, now, in each of the quadrilaterals, we draw a diagonal, then each triangle Δ_t is decomposed into certain other triangles Δ_{ts}. We shall now show that the decomposition into the triangles Δ_{ts} is for the triangles Δ_t, as well as for the triangles Δ_k, nothing else than a series of transversal

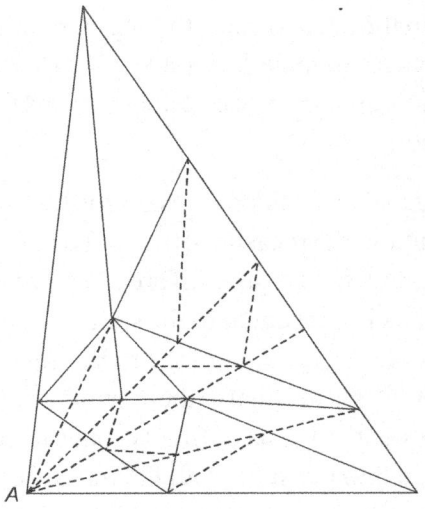

Figure 34

decompositions. In fact, it is at once evident that any decomposition of a triangle into partial triangles may always be affected by a series of transversal decompositions, providing, in this decomposition, points of division do not exist within the triangle, and further, that at least one side of the triangle remains free from points of division.

We easily see that these conditions hold for the triangles Δ_t. In fact, the interior of each of these triangles, as also one side, namely, the side opposite the point A, contains no points of division.

Likewise, for each of the triangles Δ_k, the decomposition into Δ_{ts} is reducible to transversal decompositions. Let us consider a triangle Δ_k. Among the transversals of the triangle Δ emanating from the point A, there is always a definite one to be found which either coincides with a side of Δ_k, or which itself divides Δ_k into two triangles. In the first case, the side in question always remains free from further points of division by the decomposition into the triangles Δ_{ts}. In the second case, the segment of the transversal contained within the triangle Δ_k is a side of the two triangles arising from the division, and this side certainly remains free from further points of division.

According to the considerations set forth at the beginning of this demonstration, the measure of area $F(\Delta)$ of the triangle Δ is equal to the sum of the measures of area $F(\Delta_t)$ of all the triangles Δ_t and this sum is in turn equal to the sum of all the measures of area $F(\Delta_{ts})$. However, the sum of the measures of area $F(\Delta_k)$ of all the triangles Δ_k is also equal to the sum of the measures of area $F(\Delta_{ts})$. Consequently, we have finally that the measure of area $F(\Delta)$ is also equal to the sum of all the measures of area $F(\Delta_k)$, and with this conclusion our demonstration is completed.

Definition

If we define the measure of area $F(P)$ of a polygon as the sum of the measures of area of all the triangles into which the polygon is, by a definite decomposition, divided, then upon the basis of theorem 29 and by a process of reasoning similar to that which we have employed in 18 to prove theorem 24, we know that the measure of area of a polygon is independent of the manner of decomposition into triangles and, consequently, is definitely determined by the polygon itself. From this we obtain, by aid of theorem 29, the result that *polygons of equal area have also equal measures of area.*

Furthermore, if P and Q are two polygons of equal content, then there must exist, according to the above definition, two other polygons P' and Q' of equal area, such that the polygon composed of P and P' shall be of equal area with the polygon formed by combining the polygons Q and Q'. From the two equations

$$F(P + P') = F(Q + Q')$$

$$F(P') = F(Q'),$$

we easily deduce the equation

$$F(P) = F(Q);$$

that is to say, *polygons of equal content have also equal measure of area.*

From this last statement, we obtain immediately the proof of theorem 28. If we denote the equal bases of the two triangles by g and the corresponding altitudes by h and h', respectively, then we may conclude from the assumed equality of content of the two triangles that they must also have equal measures of area; that is to say, it follows that

$$\frac{1}{2} gh = \frac{1}{2} gh'$$

and, consequently, dividing by $\dfrac{1}{2}g$, we get

$$h = h',$$

which is the statement made in theorem 28.

§21. EQUALITY OF CONTENT AND THE MEASURE OF AREA

In § 20 we have found that polygons having equal content have also equal measures of area. The converse of this is also true.

In order to prove the converse, let us consider two triangles ABC and $AB'C'$ having a common right angle at A. The measures of area of these two triangles are expressed by the formulæ

$$F(ABC) = \frac{1}{2}AB \cdot AC,$$

$$F(AB'C') = \frac{1}{2}AB' \cdot AC'.$$

We now assume that these measures of area are equal to each other, and consequently we have

$$AB \cdot AC = AB' \cdot AC',$$

or

$$AB : AB' = AC' : AC.$$

From this proposition, it follows, according to theorem 23, that the two straight lines BC' and $B'C$ are parallel, and hence, by theorem 27, the two triangles $BC'B'$ and $BC'C$ are of equal content. By the addition of the triangle ABC', it follows that the two triangles ABC and $AB'C'$ are of equal content. We have then shown that two right triangles having the same measure of area are always of equal content.

Take now any arbitrary triangle having the base g and the altitude h. Then, according to theorem 27, it has equal content with a right triangle having the two sides g and h. Since the original triangle had evidently the same measure of area as the right triangle, it follows that, in the above consideration, the restriction to right triangles was not necessary.

Hence, two *arbitrary triangles with equal measures of area are also of equal content.*

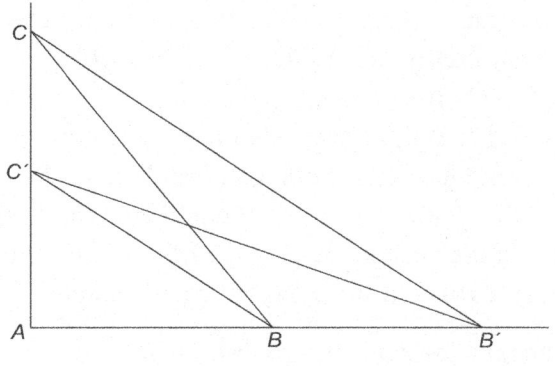

Figure 35

Moreover, let us suppose P to be any polygon having the measure of area g and let P be decomposed into n triangles having respectively the measures of area $g_1, g_2, g_3, ..., g_n$. Then, we have

$$g = g_1 + g_2 + g_3 + \cdots + g_n.$$

Construct now a triangle ABC having the base $AB = g$ and the altitude $h = 1$. Take, upon the base of this triangle, the points

$A_1, A_2, ..., A_{n-1}$ so that $g_1 = AA_1$, $g_1 = A_1A_2$, ..., $g_{n-1} = A_{n-2}A_{n-1}$, $g_n = A_{n-1}B$.

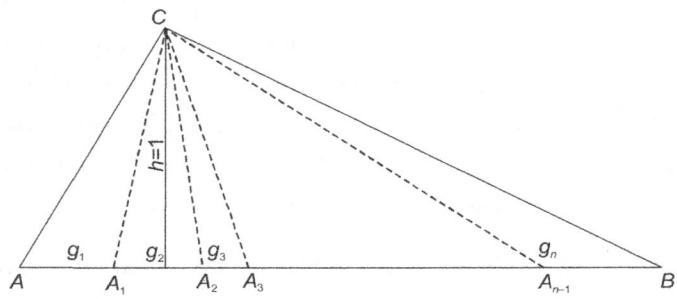

Figure 36

Since the triangles composing the polygon P have respectively the same measures of area as the triangles AA_1C, A_1A_2C, ..., A_{n-2} $A_{n-1}C$, $A_{n-1}BC$, it follows from what has already been demonstrated that they have also the same content as these triangles. Consequently, the polygon P and a triangle, having the base g and the altitude $h = 1$ are of equal content. From this, it follows, by the application of theorem 24, that two polygons having equal measures of area are always of equal content.

We can now combine the proposition of this section with that demonstrated in the last, and thus obtain the following theorem:

Theorem 30

Two polygons of equal content have always equal measures of area. Conversely, two polygons having equal measures of area are always of equal content.

In particular, if two rectangles are of equal content and have one side in common, then their remaining sides are respectively congruent. Hence, we have the following proposition:

Theorem 31

If we decompose a rectangle into several triangles by means of straight lines and then omit one of these triangles, we can no longer make up completely the rectangle from the triangles which remain.

This theorem has been demonstrated by *F.* Schur[11] and by W. Killing,[12] but by making use of the axiom of Archimedes. By O. Stolz,[13] it has been regarded as an axiom. In the foregoing discussion, it has been shown that it is completely independent of the axiom of Archimedes. However, when we disregard the axiom of Archimedes, this theorem (31) is not sufficient of itself to enable us to demonstrate Euclid's theorem concerning the equality of altitudes of triangles having equal content and equal bases. (Theorem 28.)

In the demonstration of theorems 28, 29, and 30, we have employed essentially the algebra of segments introduced in § 15, p. 29, and as this depends substantially upon Pascal's theorem (theorem 21), we see that this theorem is really the corner-stone in the theory of areas. We may, by the aid of theorems 27 and 28, easily establish the converse of Pascal's theorem.

Of two polygons P and Q, we call P the smaller or larger in content according as the measure of area $F(P)$ is less or greater than the measure of area $F(Q)$. From what has already been said, it is clear that the notions, equal content, smaller content, larger content, are mutually exclusive. Moreover, we easily see that a polygon, which lies wholly within another polygon, must always be of smaller content than the exterior one.

With this we have established the important theorems in the theory of areas.

5

DESARGUES'S THEOREM

§22. DESARGUES'S THEOREM AND ITS DEMONSTRATION FOR PLANE GEOMETRY BY AID OF THE AXIOMS OF CONGRUENCE

Of the axioms given in §§ 1–8, pp. 1–14, those of groups II–V are in part linear and in part plane axioms. Axioms 3–7 of group I are the only space axioms. In order to show clearly the significance of these axioms of space, let us assume a plane geometry and investigate, in general, the conditions for which this plane geometry may be regarded as a part of a geometry of space in which at least the axioms of groups I–III are all fulfilled.

Upon the basis of the axioms of groups I–III, it is well known that the so-called theorem of Desargues may be easily demonstrated. This theorem relates to points of intersection in a plane. Let us assume in particular that the straight line, upon which are situated the points of intersection of the homologous sides of the two triangles, is the straight line which we call the straight line at infinity. We will designate the theorem which arises in this case, together with its converse, as the theorem of Desargues. This theorem is as follows:

Theorem 32 (Desargues's Theorem)

When two triangles are so situated in a plane that their homologous sides are respectively parallel, then the lines joining the homologous vertices pass through one and the same point, or are parallel to one another.

Conversely, if two triangles are so situated in a plane that the straight lines joining the homologous vertices intersect in a common point, or are parallel to one another, and, furthermore, if two pairs of homologous sides are parallel to each other, then the third sides of the two triangles are also parallel to each other.

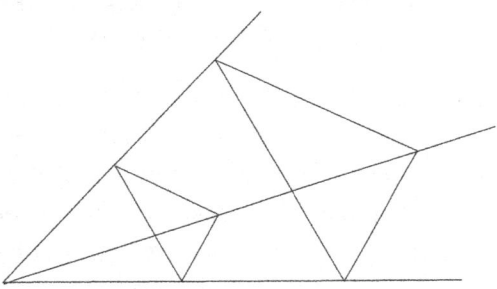

Figure 37

As we have already mentioned, theorem 32 is a consequence of the axioms I–III. Because of this fact, the validity of Desargues's theorem in the plane is, in any case, a necessary condition that the geometry of this plane may be regarded as a part of a geometry of space in which the axioms of groups I–III are all fulfilled.

Let us assume, as in §§ 13–21, pp. 23–45, that we have a plane geometry in which the axioms I, 1–2 and II–IV all hold and, also, that we have introduced in this geometry an algebra of segments conforming to § 15.

Now, as has already been established in § 17, there may be made to correspond to each point in the plane a pair of segments (x, y) and to each straight line a ratio of three segments $(u : v : w)$, so that the linear equation

$$ux + vy + w = 0$$

expresses the condition that the point is situated upon the straight line. The system composed of all the segments in our geometry forms, according to § 17, a domain of numbers for which the properties (1–16), enumerated in §13, are valid. We can, therefore, by means of this domain of numbers, construct a geometry of space in a manner similar to that already employed in § 9 or in §12, where we made use of the systems of numbers Ω and $\Omega(t)$, respectively. For this purpose, we assume that a system of three segments (x, y, z) shall represent a point, and that the ratio of four segments $(u : v : w : r)$ shall represent a plane, while a straight line is defined as the intersection of two planes. Hence, the linear equation

$$ux + vy + wz + r = 0$$

expresses the fact that the point (x, y, z) lies in the plane $(u : v : w : r)$. Finally, we determine the arrangement of the points upon a straight line, or the points of a plane with respect to a straight line situated in this plane, or the arrangement of the points in space with respect to a plane, by means of inequalities in a manner similar to the method employed for the plane in § 9.

Since we obtain again the original plane geometry by putting $z = 0$, we know that our plane geometry can be regarded as a part of geometry of space. Now, the validity of Desargues's theorem is, according to the above considerations, a necessary condition for this result. Hence, in the assumed plane geometry, it follows that Desargues's theorem must also hold.

It will be seen that the result just stated may also be deduced without difficulty from theorem 23 in the theory of proportion.

§23. THE IMPOSSIBILITY OF DEMONSTRATING DESARGUES'S THEOREM FOR THE PLANE WITHOUT THE HELP OF THE AXIOMS OF CONGRUENCE.[14]

We shall now investigate the question whether or no in plane geometry Desargues's theorem may be deduced without the assistance of the axioms of congruence. This leads us to the following result:

Theorem 33

A plane geometry exists in which the axioms I 1–2, II–III, IV 1–5, V, that is to say, all linear and all plane axioms with the exception of axiom IV, 6 of congruence, are fulfilled, but in which the theorem of Desargues (theorem 32) is not valid. Desargues's theorem is not, therefore, a consequence solely of the axioms mentioned; for, its demonstration necessitates either the space axioms or all of the axioms of congruence.

Proof Select in the ordinary plane geometry (the possibility of which has already been demonstrated in § 9, pp. 15–17) any two straight lines perpendicular to each other as the axes of x and y. Construct about the origin O of this system of co-ordinates an ellipse having the major and minor axes equal to 1 and $\frac{1}{2}$, respectively. Finally, let F denote the point situated upon the positive x-axis at the distance $\frac{3}{2}$ from O. Consider all of the circles which cut the ellipse in four real points. These points may be either distinct or in any way coincident. Of all the points situated upon these circles, we shall attempt to determine the one which lies upon the x-axis farthest from the origin. For this purpose, let us begin with an arbitrary circle cutting the ellipse in four distinct points and intersecting the positive x-axis in the point C. Suppose this circle then turned about the point C in such a manner that two or more of the four points of intersection with the ellipse finally coincide in a single point A, while the rest of them remain real. Increase now the resulting tangent circle in such a way that A always remains a point of tangency with the ellipse. In this way we obtain a circle which is either tangent to the ellipse in also a second point B, or which has with the ellipse a four-point contact in A. Moreover, this circle cuts the positive x-axis in a point more remote than C. The desired farthest point will accordingly be found among those points of intersection of the positive x-axis by circles lying exterior to the ellipse and being doubly tangent to it. All such circles must lie, as we can easily see, symmetrically with respect to the y-axis. Let a, b be the co-ordinates of any point on the ellipse. Then an easy calculation shows that the circle, which is symmetrical with respect to y-axis and tangent to the ellipse at this point, cuts off from the positive x-axis the segment .

$$x = |\sqrt{1+3b^2}|$$

The greatest possible value of this expression occurs for $b = \frac{1}{2}$ and, hence, is equal to $\frac{1}{2}|\sqrt{7}|$. Since the point on the x-axis which we have denoted by F has for its abscissa the value $\frac{3}{2} > \frac{1}{2}|\sqrt{7}|$, it follows that *among the circles cutting the ellipse four times there is certainly none which passes through the point F.*

We will now construct a new plane geometry in the following manner. As points in this new geometry, let us take the points of the (xy)-plane. We will define a straight line of our new geometry in the following manner. Every straight line of the (xy)-plane which is either tangent to the fixed ellipse, or does not cut it at all, is taken unchanged as a straight line of the new geometry. However, when any straight line g of the (xy)-plane cuts the ellipse, say in the points P and Q, we will then define the corresponding straight line of the new geometry as follows. Construct a circle passing through the points P and Q and the fixed point F. From what has just been said, this circle will have no other point in common with the ellipse. We will now take the broken line, consisting of the arc PQ just mentioned and the two parts of the straight line g extending outward indefinitely from the points P and Q, as the

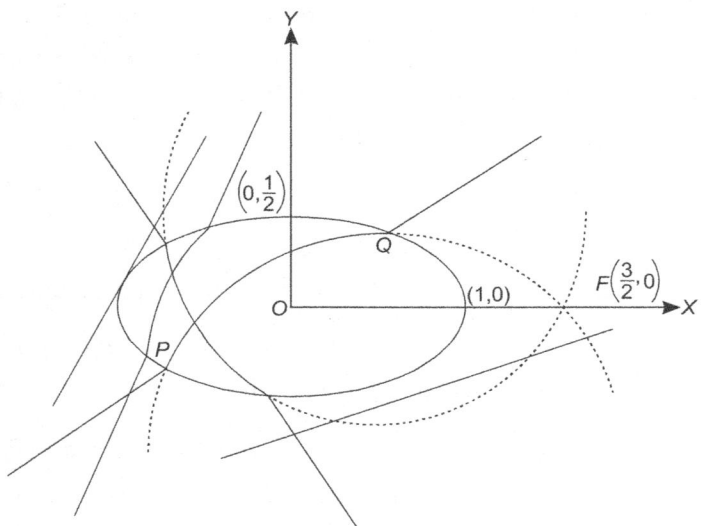

Figure 38

required straight line in our new geometry. Let us suppose all of the broken lines constructed which correspond to straight lines of the (xy)-plane. We have then a system of broken lines which, considered as straight lines of our new geometry, evidently satisfy the axioms I, 1–2 and III. By a convention as to the actual arrangement of the points and the straight lines in our new geometry, we have also the axioms II fulfilled.

Moreover, we will call two segments AB and $A'B'$ congruent in this new geometry, if the broken line extending between A and B has equal length, in the ordinary sense of the word, with the broken line extending from A' to B'.

Finally, we need a convention concerning the congruence of angles. So long as neither of the vertices of the angles to be compared lies upon the ellipse, we call the two angles congruent to each other, if they are equal in the ordinary sense. In all other cases we make the following convention. Let A, B, C be points which follow one another in this order upon a straight line of our new geometry, and let A', B', C' be also points which lie in this order upon another straight line of our new geometry. Let D be a point lying outside of the straight line ABC and D' be a point outside of the straight A', B', C'. We will now say that, in our new geometry, the angles between these straight lines fulfill the congruences

$$\angle ABD \equiv \angle A'B'D' \text{ and } \angle CBD = \angle C'B'D'$$

whenever the natural angles between the corresponding broken lines of the ordinary geometry fulfill the proportion

$$\angle ABD : \angle CBD = \angle A'B'D' : \angle C'B'D'.$$

These conventions render the axioms IV, 1–5 and V valid.

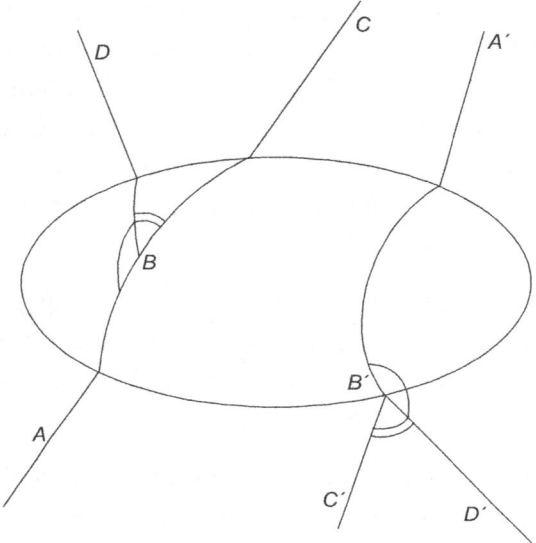

Figure 39

In order to see that Desargues's theorem does not hold for our new geometry, let us consider the following three ordinary straight lines of the (xy)-plane; viz., the axis of x, the axis of y, and the straight line joining the two points of the ellipse $\left(\dfrac{3}{5}, \dfrac{2}{5}\right)$ and $\left(-\dfrac{3}{5}, -\dfrac{2}{5}\right)$. Since these three ordinary straight lines pass through the origin, we can easily construct two triangles so that their vertices shall lie respectively upon these three straight lines and their homologous sides shall be parallel and all three sides shall lie exterior to the ellipse. As we may see from figure 40, or show by an easy calculation, the broken lines arising from the three straight lines in question do not intersect in a common point. Hence, it follows that Desargues's theorem certainly does not hold for this particular plane geometry in which we have constructed the two triangles just considered.

This new geometry serves at the same time as an example of a plane geometry in which the axioms I, 1–2, II–III, IV, 1–5, V all hold, but which cannot be considered as a part of a geometry of space.

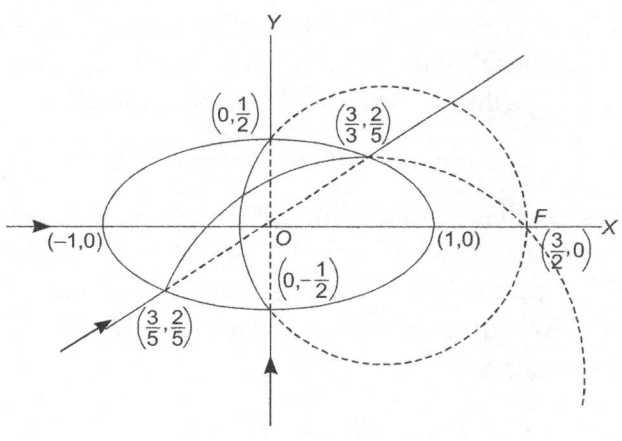

Figure 40

§24. INTRODUCTION OF AN ALGEBRA OF SEGMENTS BASED UPON DESARGUES'S THEOREM AND INDEPENDENT OF THE AXIOMS OF CONGRUENCE.[15]

In order to see fully the significance of Desargues's theorem (theorem 32), let us take as the basis of our consideration a plane geometry where all of the axioms I 1–2, II–III are valid, that is to say, where all of the plane axioms of the first three groups hold, and then introduce into this geometry, in the following manner, a new algebra of segments independent of the axioms of congruence.

Take in the plane two fixed straight lines intersecting in O, and consider only such segments as have O for their origin and their other extremity in one of the fixed lines. We will regard the point O itself as a segment and call it the segment O. We will indicate this fact by writing

$$OO = 0, \text{ or } 0 = OO.$$

Let E and E' be two definite points situated respectively upon the two fixed straight lines through O. Then, define the two segments OE and OE' as the segment 1 and write accordingly

$$OE = OE' = 1 \quad \text{or} \quad 1 = OE = OE'. \tag{1}$$

We will call the straight line EE', for brevity, the unit-line. If, furthermore, A and A' are points upon the straight lines OE and OE', respectively, and, if the straight line AA' joining them is parallel to EE', then we will say that the segments OA and OA' are equal to one another, and write

$$OA = OA' \quad \text{or} \quad OA' = OA.$$

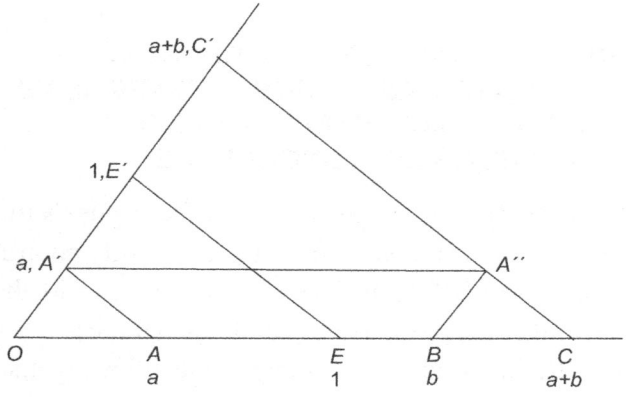

Figure 41

In order now to define the sum of the segments $a = OA$ and $b = OB$, we construct AA' parallel to the unit-line EE' and draw through A' a parallel to OE and through B a parallel to OE'. Let these two parallels intersect in A''. Finally, draw through A'' a straight line parallel to the unit-line EE'. Let this parallel cut the two fixed lines OE and OE' in C and C' respectively. Then $c = OC = OC'$ is called the *sum* of the segments $a = OA$ and $b = OB$. We indicate this by writing

$$c = a + b, \text{ or } a + b = c$$

In order to define the product of a segment $a = OA$ by a segment $b = OB$, we make use of exactly the same construction as employed in § 15, except that, in place of the sides of a right angle, we make use here of the straight lines OE and OE'. The construction is consequently as follows. Determine upon OE' a point A' so that AA' is parallel to the unit-line EE', and join E with A'. Then draw through B a straight line parallel to EA'.

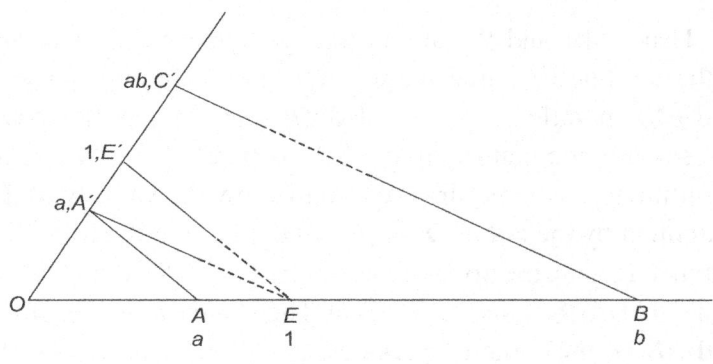

Figure 42

This parallel will intersect the fixed straight line OE' in the point C', and we call $c = OC'$ the product of the segment $a = OA$ by the segment $b = OB$. We indicate this relation by writing

$$c = ab, \text{ or } ab = c.$$

§25. THE COMMUTATIVE AND THE ASSOCIATIVE LAW OF ADDITION FOR OUR NEW ALGEBRA OF SEGMENTS

In this section, we shall investigate the laws of operation, as enumerated in § 13, in order to see which of these hold for our new algebra of segments, when we base our considerations upon a plane geometry in which axioms I 1–2, II–III are all fulfilled, and, moreover, in which Desargues's theorem also holds.

First of all, we shall show that, for the addition of segments as defined in § 24, the commutative law

$$a + b = b + a$$

holds. Let

$$a = OA = OA'$$
$$b = OB = OB'$$

Hence, AA' and BB' are, according to our convention, parallel to the unit-line. Construct the points A'' and B'' by drawing $A'A''$ and $B'B''$ parallel to OA and also AB' and BA' parallel to OA. We see at once that the line $A''B''$ is parallel to AA' as the commutative law requires. We shall show the validity of this statement by the aid of Desargues's theorem in the following manner. Denote the point of intersection of AB'' and $A'A''$ by F and that of BA'' and $B'B''$ by D. Then, in the triangles $AA'F$ and $BB'D$, the homologous sides are parallel to each other. By Desargues's theorem, it follows that the three points O, F, D lie in a straight line. In consequence of this condition, the two triangles OAA' and $DB''A''$ lie in such a way that the lines joining the corresponding vertices pass through the same point F, and since the homologous sides OA and DB'', as also OA' and DA'', are parallel to each other, then, according to the second part of

Desargues's theorem (theorem 32), the third sides AA' and $B''A''$ are parallel to each other.

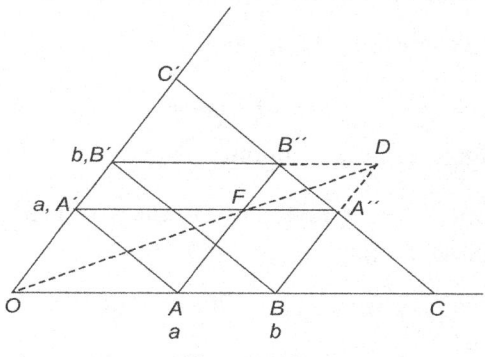

Figure 43

To prove the associative law of addition

$$a + (b + c) = (a + b) + c,$$

we shall make use of figure 44. In consequence of the commutative law of addition just demonstrated, the above formula states that the straight line $A''B''$ must be parallel to the unit-line. The validity of this statement is evident, since the shaded part of figure 44 corresponds exactly with figure 43.

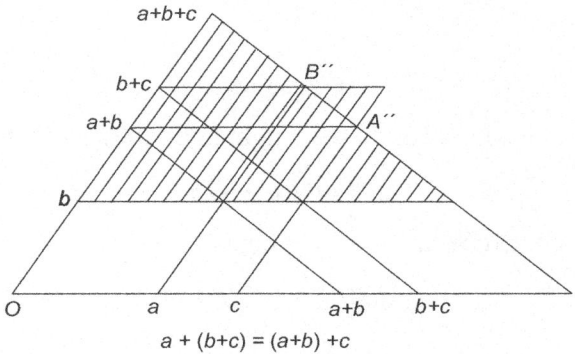

$$a + (b+c) = (a+b) + c$$

Figure 44

§26. THE ASSOCIATIVE LAW OF MULTIPLICATION AND THE TWO DISTRIBUTIVE LAWS FOR THE NEW ALGEBRA OF SEGMENTS

The associative law of multiplication

$$a(bc) = (ab)c$$

has also a place in our new algebra of segments.

Let there be given upon the first of the two fixed straight lines through O the segments

$$1 = OA, \quad b = OC, \quad c = OA'$$

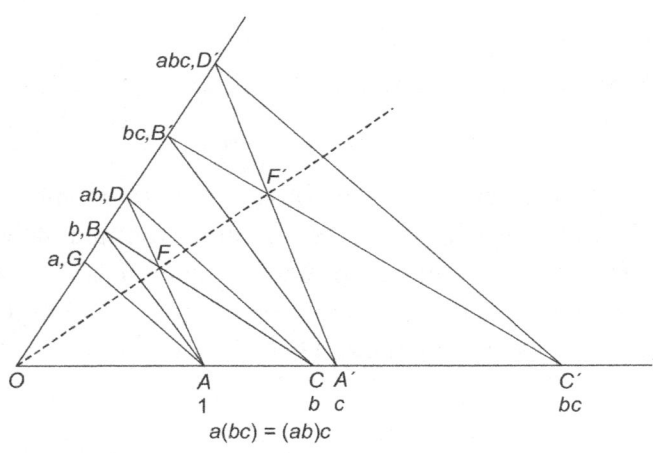

Figure 45

and upon the second of these straight lines, the segments

$$a = OG, \; b = OB.$$

In order to construct the segments

$$bc = OB', \text{ and } bc = OC',$$
$$ab = OD,$$
$$(ab)c = OD',$$

in accordance with §24, draw $A'B'$ parallel to AB, $B'C'$ parallel to BC, CD parallel to AG, and $A'D'$ parallel to AD. We see at once that the given law amounts to the same as saying that CD must also be parallel to $C'D'$. Denote the point of intersection of the straight lines $A'D'$ and $B'C'$ by F' and that of the straight lines AD and BC by F. Then the triangles ABF and $A'B'F'$ have their homologous sides parallel to each other, and, according to Desargues's theorem, the three points O, F, F' must lie in a straight line. Because of these conditions, we can apply the second part of Desargues's theorem to the two triangle CDF and $C'D'F'$, and hence show that, in fact, CD is parallel to $C'D'$.

Finally, upon the basis of Desargues's theorem, we shall show that the two distributive laws

$$a(b + c) = ab + ac$$

and

$$(a + b)c = ac + bc$$

hold for our algebra of segments.

In the proof of the first one of these laws, we shall make use of figure 46.[16] In this figure, we have

$$b = OA', c = OC',$$

$$ab = OB', \ ab = OA'', \ ac = OC'', \text{ etc.}$$

In the same figure, $B''D_2$ is parallel to $C''D_1$ which is parallel to the fixed straight line OA', and $B'D_1$ is parallel to $C'D_2$, which is parallel to the fixed straight line OA''. Moreover, we have $A'A''$ parallel to $C'C''$, and $A'B''$ parallel to $B'A''$, parallel to $F'D_2$, parallel to $F''D_1$.

Our proposition amounts to asserting that we must necessarily have also

$F'F''$ parallel to $A'A''$ and to $C'C''$.

We construct the following auxiliary lines:

$F''J$ parallel to the fixed straight line OA',

$F'J$ parallel to the fixed straight line OA''.

Let us denote the points of intersection of the straight lines $C''D_1$ and $C'D_2$, $C''D_1$ and $F'J$, $C'D_2$ and $F''J$ by G, H_1, H_2, respectively. Finally, we obtain the other auxiliary lines indicated in the figure by joining the points already constructed.

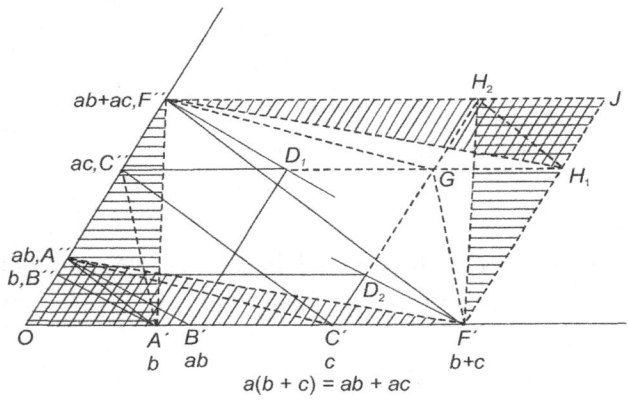

Figure 46

In the two triangles $A'B''C''$ and $F'D_2G$, the straight lines joining homologous vertices are parallel to each other. According to the second part of Desargues's theorem, it follows, therefore, that

$A'C''$ is parallel to $F'G$.

In the two triangles $A'C''F''$ and $F'GH_2$, the straight lines joining the homologous vertices are also parallel to each other. From the properties already demonstrated, it follows by virtue of the second part of Desargues's theorem that we must have

$$A'F'' \text{ parallel to } F'H_2.$$

Since in the two horizontally shaded triangles $OA'F'$ and JH_2F' the homologous sides are parallel, Desargues's theorem shows that the three straight lines joining the homologous vertices, viz.:

$$OJ, \ A'H_2, F''F' \tag{2}$$

all intersect in one and the same point, say in P.

In the same way, we have necessarily

$$A''F' \text{ parallel to } F''H_1$$

and since, in the two obliquely shaded triangles $OA''F'$ and JH_1F'', the homologous sides are parallel, then, in consequence of Desargues's theorem, the three straight lines joining the homologous vertices, viz.:

$$OJ, \ A''H_1, F'F''$$

all intersect likewise in the same point, namely, in point P.

Moreover, in the triangles $OA'A''$ and JH_2H_1, the straight lines joining the homologous vertices all pass through this same point P, and, consequently, it follows that we have

$$H_1H_2 \text{ parallel to } A'A'',$$

and, therefore,

$$H_1H_2 \text{ parallel to } C'C'',$$

Finally, let us consider the figure $F''H_2C'C''H_1F'F''$. Since, in this figure, we have

$$F''H_2 \text{ parallel to } C'F', \text{ parallel to } C''H_1,$$

$$GH_2C' \text{ parallel to } F''C'', \text{ parallel to } H_1F',$$

$$C'C'' \text{ parallel to } H_1H_2,$$

we recognize here again figure 43, which we have already made use of in § 25 to prove the commutative law of addition. The conclusions, analogous to those which we reached there, show that we must have

$$F'F'' \text{ parallel to } H_1H_2$$

and, consequently, we must have also

$$F'F'' \text{ parallel to } A'A'',$$

which result concludes our demonstration.

To prove the second formula of the distributive law, we make use of an entirely different figure,—figure 47. In this figure, we have

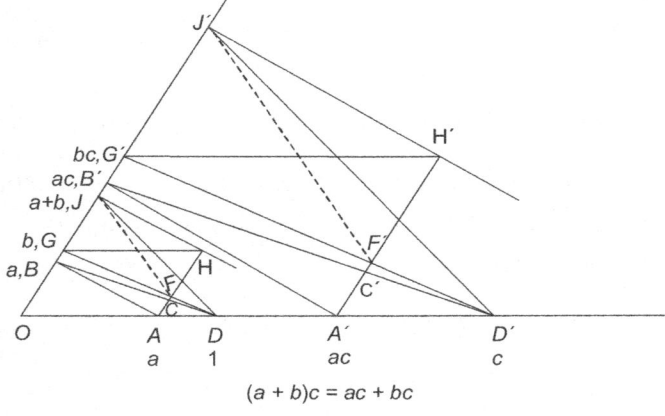

$$(a + b)c = ac + bc$$

Figure 47

$$1 = OD, a = OA, a = OB, b = OG, c = OD',$$
$$ac = OA', ac = OB', BC = OG', \text{etc.},$$

and, furthermore, we have

GH parallel to $G'H'$, parallel to the fixed line OA,

GH parallel to $A'H'$, parallel to the fixed line OB,

We have also

AB parallel to $A'B'$

BD parallel to $B'D'$

DG parallel to $D'G'$

HJ parallel to $H'J'$.

That which we are to prove amounts, then, to demonstrating that

DJ must be parallel to $D'J'$.

Denote the points in which BD and GD intersect the straight line AH by C and F, respectively, and the points in which $B'D'$ and $G'D'$ intersect the straight line $A'H'$ by C' and F', respectively. Finally, draw the auxiliary lines FJ and $F'J'$, indicated in the figure by dotted lines.

In the triangles ABC and $A'B'C'$, the homologous sides are parallel and, consequently, by Desargues's theorem the three points O, C, C' lie on a straight line. Then, by considering in the same way the triangles CDF and $CD'F'$, it follows that the points O, F, F' lie upon the same straight line and likewise, from a consideration of the triangles FGH and $F'H'J'$, we find the points O, H, H' to be situated on a straight line. Now, in the triangles FHJ and $F'H'J'$, the straight lines joining the homologous vertices all pass through the same point O, and, hence, as a consequence of the second part of Desargues's theorem, the straight lines FJ and $F'J'$ must also be parallel to each other. Finally, a consideration of the triangles DFJ and $D'F'J'$ shows that the straight lines DJ and $D'J'$ are parallel to each other and with this our proof is completed.

§27. EQUATION OF THE STRAIGHT LINE, BASED UPON THE NEW ALGEBRA OF SEGMENTS

In §§ 24–26, we have introduced into the plane geometry an algebra of segments in which the commutative law of addition and that of multiplication, as well as the two distributive laws, hold. This was done upon the assumption that the axioms cited in § 24, as also the theorem of Desargues, were valid. In this section, we shall show how an analytical representation of the point and straight line in the plane is possible upon the basis of this algebra of segments.

Definition

Take the two given fixed straight lines lying in the plane and intersecting in O as the axis of x and of y, respectively. Let us suppose any point P of the plane determined by the two segments x, y which we obtain upon the x-axis and y-axis, respectively, by drawing through P parallels to these axes. These segments are called the *co-ordinates* of the point P. Upon the basis of this new algebra of segments and by aid of Desargues's theorem, we shall deduce the following proposition.

Theorem 34

The co-ordinates x, y of a point on an arbitrary straight line always satisfy an equation in these segments of the form

$$ax + by + c = 0.$$

In this equation, the segments *a* and *b* stand necessarily to the left of the co-ordinates *x* and *y*. The segments *a* and *b* are never both zero and *c* is an arbitrary segment.

Conversely, every equation in these segments and of this form represents always a straight line in the plane geometry under consideration.

Proof Suppose that the straight line l passes through the origin *O*. Furthermore, let *C* be a definite point upon *l* different from *O*, and *P* any arbitrary point of *l*. Let *OA* and *OB* be the co-ordinates of *C* and *x, y* be the co-ordinates of *P*. We will denote the straight line joining the extremities of the segments *x, y* by *g*. Finally, through the extremity of the segment 1, laid off on the *x*-axis, draw a straight line *h* parallel to *AB*.

This parallel cuts off upon the *y*-axis the segment *e*. From the second part of Desargues's theorem, it follows that the straight line *g* is also always parallel to *AB*. Since *g* is always parallel to *h*,

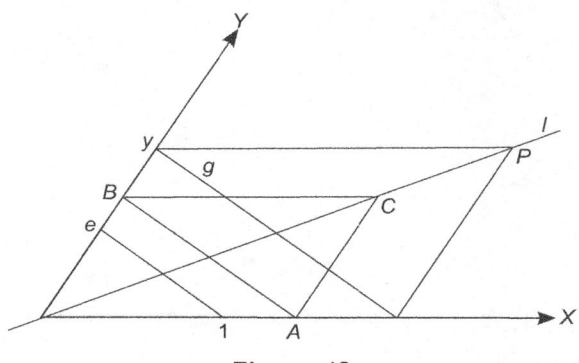

Figure 48

it follows that the co-ordinates *x, y* of the point *P* must satisfy the equation

$$ex = g.$$

Moreover, in figure 49 let l' be any arbitrary straight line in our plane. This straight line will cut off on the x-axis the segment $c = OO'$. Now, in the same figure, draw through O the straight line l parallel to l'. Let P' be an arbitrary point on the line l'. The straight line through P', parallel to the x-axis, intersects the straight line l in P and cuts off upon the y-axis the segment $y = OB$. Finally, through P and P' let parallels to the y-axis cut off on the x-axis the segments $x = OA$ and $x' = OA'$.

We shall now undertake to show that the equation

$$x' = x + c$$

is fulfilled by the segments in question. For this purpose, draw $O'C$ parallel to the unit-line and likewise CD parallel to the x-axis and AD parallel to the y-axis.

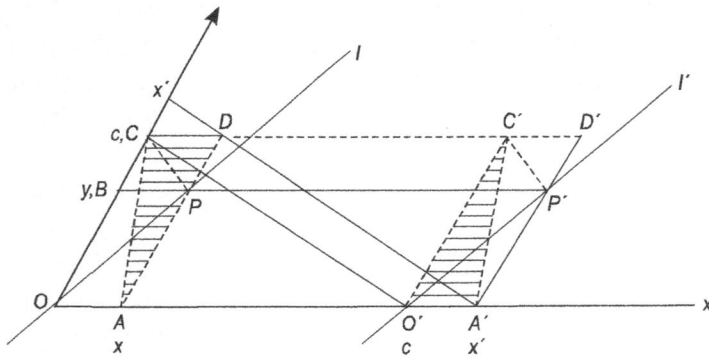

Figure 49

Then, to prove our proposition amounts to showing that we must have necessarily

$$A'D \text{ parallel to } O'C.$$

Let D' be the point of intersection of the straight lines CD and $A'P'$ and draw $O'C'$ parallel to the y-axis.

Since, in the triangles OCP and $O'C'P'$, the straight lines joining the homologous vertices are parallel, it follows, by virtue of the second part of Desargues's theorem, that we must have

CP parallel to $C'P'$.

In a similar way, a consideration of the triangles ACP and $A'C'P'$ shows that we must have

AC parallel to $A'C'$.

Since, in the triangles ACD and $C'A'O'$, the homologous sides are parallel to each other, it follows that the straight lines AC', CA' and DO' intersect in a common point. A consideration of the triangles $C'A'D$ and ACO' then shows that $A'D$ and CO' are parallel to each other. From the two equations already obtained, viz.:

$$ex = y \text{ and } x' = x+c, \tag{3}$$

follows at once the equation

$$ex' = y + ec. \tag{4}$$

If we denote, finally, by n the segment which added to the segment 1 gives the segment 0, then, from this last equation, we may easily deduce the following

$$ex' + ny + nec = 0, \tag{5}$$

and this equation is of the form required by theorem 34.

We can now show that the second part of the theorem is equally true; for, every linear equation

$$ax + by + c = 0 \tag{6}$$

may evidently be brought into the required form

$$ex + ny + nec = 0 \qquad (7)$$

by a left-sided multiplication by a properly chosen segment.

It must be expressly stated, however, that, by our hypothesis, an equation of segments of the form

$$xa + yb + c = 0, \qquad (8)$$

where the segments a, b stand to the right of the co-ordinates x, y does not, in general, represent a straight line.

In Section 30, we shall make an important application of theorem 34.

§28. THE TOTALITY OF SEGMENTS, REGARDED AS A COMPLEX NUMBER SYSTEM

We see immediately that, for the new algebra of segments established in Section 24, theorems 1–6 of Section 13 are fulfilled. Moreover, by aid of Desargues's theorem, we have already shown in Sections 25 and 26 that the laws 7–11 of operation, as given in Section 13, are all valid in this algebra of segments. With the single exception of the commutative law of multiplication, therefore, all of the theorems of connection hold.

Finally, in order to make possible an order of magnitude of these segments, we make the following convention. Let A and B be any two distinct points of the straight line OE. Suppose then that the four points O, E, A, B stand, in conformity with axiom II, 4, in a certain sequence. If this sequence is one of the following six possible ones, viz.:

ABOE, AOBE, AOEB, OABE, OAEB, OEAB,

then we will call the segment $a = OA$ *smaller* than the segment $b = OB$ and indicate the same by writing

$$a < b.$$

On the other hand, if the sequence is one of the six following ones, viz.:

$$BAOE, \; BOAE, \; BOEA, \; OBAE, \; OBEA, \; OEBA,$$

then we will call the segment $a = OA$ greater than the segment $b = OB$, and we write accordingly

$$a > b.$$

This convention remains in force whenever A or B coincides with O or E, only then the coinciding points are to be regarded as a single point, and, consequently, we have only to consider the order of three points.

Upon the basis of the axioms of group II, we can easily show also that, in our algebra of segments, the laws 13–16 of operation given in Section 13 are fulfilled. Consequently, the totality of all the different segments forms a complex number system for which the laws 1–11, 13–16 of Section 13 hold; that is to say, all of the usual laws of operation except the commutative law of multiplication and the theorem of Archimedes. We will call such a system, briefly, a *desarguesian number system*.

§29. CONSTRUCTION OF A GEOMETRY OF SPACE BY AID OF A DESARGUESIAN NUMBER SYSTEM

Suppose we have given a desarguesian number system D. Such a system makes possible the construction of a geometry of space in which axioms I, II, III are all fulfilled.

In order to show this, let us consider any system of three numbers (x, y, z) of the desarguesian number system D as a point, and the ratio of four such numbers $(u : v : w : r)$, of which the first three are not 0, as a plane. However, the systems $(u : v : w : r)$ and $(au : av : aw : ar)$, where a is any number of D different from 0, represent the same plane. The existence of the equation

$$ux + vy + wz + r = 0$$

expresses the condition that the point (x, y, z) shall lie in the plane $(u : v : w : r)$. Finally, we define a straight line by the aid of a system of two planes $(u' : v' : w' : r')$ and $(u'' : v'' : w'' : r'')$, where we impose the condition that it is impossible to find in D two numbers a', a'' different from zero, such that we have simultaneously the relations

$$a'u' = a''u'', a'v' = a''v'', a'w' = a''w'',$$

A point (x, y, z) is said to be situated upon this straight line $[(u' : v' : w' : r'), (u'' : v'' : w'' : r'')]$, if it is common to the two planes $(u' : v' : w' : r')$ and $(u'' : v'' : w'' : r'')$. Two straight lines which contain the same points are not regarded as being distinct.

By application of the laws 1–11 of § 13, which by hypothesis hold for the numbers of D, we obtain without difficulty the result that the geometry of space which we have just constructed satisfies all of the axioms of groups I and III.

In order that the axioms (II) of order may also be valid, we adopt the following conventions. Let

$$(x_1, y_1, z_1), (x_2, y_2, z_2), (x_3, y_3, z_3)$$

be any three points of a straight line

$$[(u' : v' : w' : r'), (u'' : v'' : w'' : r'')].$$

Then, the point (x_2, y_2, z_2) is said to lie between the other two, if we have fulfilled at least one of the six following double inequalities:

$$x_1 < x_2 < x_3, \quad x_1 > x_2 > x_3, \tag{9}$$

$$y_1 < y_2 < y_3, \quad y_1 > y_2 > y_3, \tag{10}$$

$$z_1 < z_2 < z_3, \quad z_1 > z_2 > z_3, \tag{11}$$

If one of the two double inequalities (1) exists, then we can easily conclude that either $y_1 = y_2 = y_3$ or one of the two double inequalities (2) exists, and, consequently, either $z_1 = z_2 = z_3$ or one of the double inequalities (3) must exist. In fact, from the equations

$$u'x_i + v'y_i + w'z_i + r' = 0,$$
$$u''x_i + v''y_i + w''zi + r'' = 0,$$
$$(i = 1, 2, 3)$$

we may obtain, by a left-sided multiplication of these equations by numbers suitably chosen from D and then adding the resulting equations, a system of equations of the form

$$u'''x_i + v'''y_i + r''' = 0, \quad (i = 1, 2, 3). \tag{12}$$

In this system, the coefficient v''' is certainly different from zero, since otherwise the three numbers x_1, x_2, x_3 would be mutually equal.

From

$$x_1 \lessgtr x_2 \lessgtr x_3,$$

it follows that

$$u'''x_1 \lesseqgtr x'''u_2 \lesseqgtr x'''u_3,$$

and, hence, as a consequence of (4), we have

$$v''' y_1 + r''' \gtreqless v''' y_2 + r''' \gtreqless v''' y_3 + r'''$$

and, therefore,

$$v''' y_1 \gtreqless v''' y_2 \gtreqless v''' y_3.$$

Since v''' is different from zero, we have

$$y_1 \gtreqless y_2 \gtreqless y_3.$$

In each of these double inequalities, we must take either the upper sign throughout, or the middle sign throughout, or the lower sign throughout.

The preceding considerations show, that, in our geometry, the linear axioms II, 1–4 of order are all valid. However, it remains yet to show that, in this geometry, the plane axiom II, 5 is also valid.

For this purpose let a plane $(u : v : w : r)$ and a straight line $[(u : v : w : r), (u' : v' : w' : r')]$ in this plane be given. Let us assume that all the points (x, y, z) of the plane $(u : v : w : r)$, for which we have the expression $u'x + v'y + w'z + r'$ greater than or less than zero, lie respectively upon the one side or upon the other side of the given straight line. We have then only to show that this convention is in accordance with the preceding statements. This, however, is easily done. We have thus shown that all of the axioms of groups I, II, III are fulfilled in the geometry of space which we have obtained in the above indicated manner from the desarguesian number system D. Remembering now that the theorem of Desargues is a consequence of the axioms I, II, III, we see that the proposition just stated is exactly the converse of the result reached in Section 28.

§30. SIGNIFICANCE OF DESARGUES'S THEOREM

If, in a plane geometry, axioms I, 1–2, II, III are all fulfilled and, moreover, if the theorem of Desargues holds, then, according to §§ 24–28, it is always possible to introduce into this geometry an algebra of segments to which the laws 1–11, 13–16 of § 13 are applicable. We will now consider the totality of these segments as a complex number system and construct, upon the basis of this system, a geometry of space, in accordance with § 29, in which all of the axioms I, II, III hold.

In this geometry of space, we shall consider only the points $(x, y, 0)$ and those straight lines upon which only such points lie. We have then a plane geometry which must, if we take into account the proposition established in § 27, coincide exactly with the plane geometry proposed at the beginning. Hence, we are led to the following proposition, which may be regarded as the objective point of the entire discussion of the present chapter.

Theorem 35

If, in a plane geometry, axioms I, 1–2, II, III are all fulfilled, then the existence of Desargues's theorem is the necessary and sufficient condition that this plane geometry may be regarded as a part of a geometry of space in which all of the axioms I, II, III are fulfilled.

The theorem of Desargues may be characterized for plane geometry as being, so to speak, the result of the elimination of the space axioms.

The results obtained so far put us now in the position to show that every geometry of space in which axioms I, II, III are all fulfilled may be always regarded as a part of a "geometry of any number of dimensions whatever." By a geometry of an arbitrary number of dimensions is to be understood the totality of all points, straight lines, planes, and other linear elements, for which the corresponding axioms of connection and of order, as well as the axiom of parallels, are all valid.

6

PASCAL'S THEOREM

§31. TWO THEOREMS CONCERNING THE POSSIBILITY OF PROVING PASCAL'S THEOREM

As is well known, Desargues's theorem (theorem 32) may be demonstrated by the aid of axioms I, II, III; that is to say, by the use, essentially, of the axioms of space. In § 23, we have shown that the demonstration of this theorem without the aid of the space axioms of group I and without the axioms of congruence (group IV) is impossible, even if we make use of the axiom of Archimedes.

Upon the basis of axioms I, 1–2, II, III, IV and, hence, by the exclusion of the axioms of space but with the assistance, essentially, of the axioms of congruence, we have, in § 14, deduced Pascal's theorem and, consequently, according to § 22, also Desargues's theorem. The question arises as to whether Pascal's theorem can be demonstrated without the assistance of the axioms of congruence. Our investigation will show that in this respect Pascal's theorem is very different from Desargues's theorem; for, in the demonstration of Pascal's theorem, the admission or exclusion of the axiom of Archimedes is of decided influence. We may combine the essential results of our investigation in the two following theorems.

Theorem 36

Pascal's theorem (theorem 21) may be demonstrated by means of the axioms I, II, III, V; that is to say, without the assistance of the axioms of congruence and with the aid of the axiom of Archimedes.

Theorem 37

Pascal's theorem (theorem 21) cannot be demonstrated by means of the axioms I, II, III alone; that is to say, by exclusion of the axioms of congruence and also the axiom of Archimedes.

In the statement of these two theorems, we may, by virtue of the general theorem 35, replace the space axioms I, 3–7 by the plane condition that Desargues's theorem (theorem 32) shall be valid.

§32. THE COMMUTATIVE LAW OF MULTIPLICATION FOR AN ARCHIMEDEAN NUMBER SYSTEM

The demonstration of theorems 36 and 37 rests essentially upon certain mutual relations concerning the laws of operation and the fundamental propositions of arithmetic, a knowledge of which is of itself of interest. We will state the two following theorems.

Theorem 38

For an archimedean number system, the commutative law of multiplication is a necessary consequence of the remaining laws of operation; that is to say, if a number system possesses the properties

1–11, 13–17 given in § 13, it follows necessarily that this system satisfies also formula 12.

Proof   Let us observe first of all that, if a is an arbitrary number of the system, and, if

$$n = 1 + 1 + \cdots + 1$$

is a positive integral rational number, then for n and a the commutative law of multiplication always holds. In fact, we have

$$an = a(1 + 1 + \cdots + 1)$$
$$= a \cdot 1 + a \cdot 1 + \cdots + a \cdot 1$$
$$= a + a + \cdots + a$$

and likewise

$$na = (1 + 1 + \cdots + 1)a$$
$$= 1 \cdot a + 1 \cdot a + \cdots + 1 \cdot a$$
$$= a + a + \cdots + a.$$

Suppose now, in contradiction to our hypothesis, a, b to be numbers of this system, for which the commutative law of multiplication does not hold. It is then at once evident that we may make the assumption that we have

$$a > 0,\ b > 0,\ ab - ba > 0.$$

By virtue of condition 6 of § 13, there exists a number $c(> 0)$, such that

$$(a + b + 1)c = ab - ba.$$

Finally, if we select a number d, satisfying simultaneously the inequalities

$$d > 0,\ d < 1,\ d < c,$$

and denote by m and n two such integral rational numbers ≥ 0 that we have respectively

$$md < a \leq (m + 1)d$$

and

$$nd < b \leq (n + 1)d$$

then the existence of the numbers m and n is an immediate consequence of the theorem of Archimedes (theorem 17, § 13). Recalling now the remark made at the beginning of this proof, we have by the multiplication of the last inequalities

$$ab \leqq mnd^2 + (m + n + 1)d^2$$

$$ba > mnd^2,$$

and, hence, by subtraction

$$ab - ba \leqq (m + n + 1)d^2.$$

We have, however,

$$md < a, \ nd < b, \ , \ d < 1$$

and, consequently,

$$(m + n + 1)d < a + b + 1;$$

i.e.,

$$ab - ba < (a + b + 1)d,$$

or, since $d < c$, we have

$$ab - ba < (a + b + 1)c.$$

This inequality stands in contradiction to the definition of the number c, and, hence, the validity of the theorem 38 follows.

§33. THE COMMUTATIVE LAW OF MULTIPLICATION FOR A NON-ARCHIMEDEAN NUMBER SYSTEM

Theorem 39

For a non-archimedean number system, the commutative law of multiplication is not a necessary consequence of the remaining laws of operation; that is to say, there exists a system of numbers possessing the properties 1–11, 13–16 mentioned in § 13, but for which the commutative law (12) of multiplication is not valid. A desarguesian number system, in the sense of § 28, is such a system.

Proof   Let t be a parameter and T any expression containing a finite or infinite number of terms, say of the form

$$T = r_0 t^n + r_1 t^{n+3} + r_2 t^{n+2} + r_3 t^{n+3} + \ldots,$$

where $r_0 (\neq 60), r_1, r_2$ ... are arbitrary rational numbers and n is an arbitrary integral rational number $\gtreqless 0$. Moreover, let s be another parameter and S any expression having a finite or infinite number of terms, say of the form

$$S = s^m T_0 + s^{m+1} T_1 + s^{m+2} T_2 + \ldots,$$

where $T_0 (\neq 60), T_1, T_2, \ldots$ denote arbitrary expressions of the form T and m is again an arbitrary integral rational number $\gtreqless 0$. We will regard the totality of all the expressions of the form S as a complex number system $\Omega(s, t)$, for which we will assume the following laws of operation; namely, we will operate with s and t according to the laws 7–11 of § 13, as with parameters, while in place of rule 12 we will apply the formula

$$ts = 2st. \tag{13}$$

If, now, S', S'' are any two expressions of the form S, say

$$S' = sm'T_0' + s^{m'+1}T_1' + s^{m'+2}T_2' \cdots,$$

$$S'' = s^{m''}T_0'' + s^{m''+1}T_1'' + s^{m''+2}T_2'' + \cdots,$$

then, by combination, we can evidently form a new expression $S' + S''$ which is of the form S, and is, moreover, uniquely determined. This expression $S' + S''$ is called the sum of the numbers represented by S' and S''.

By the multiplication of the two expressions S' and S'' term by term, we obtain another expression of the form

$$S'S'' = sm'T_0'sm''T_0'' + (sm'T_0's^{m''+1}T_1'' - s^{m'+1}T_1'sm''T_0'')$$

$$+ (sm'T_0's^{m''+2}T_2'' + s^{m'+1}T_1's^{m''+1}T_1'' + s^{m'+2}T_2'sm''T_0'') + \cdots$$

This expression, by the aid of formula (1), is evidently a definite single-valued expression of the form S and we will call it the product of the numbers represented by S' and S''.

This method of calculation shows at once the validity of the laws 1–5 given in Section 13 for calculating with numbers. The validity of law 6 of that section is also not difficult to establish. To this end, let us assume that

$$S' = s^{m'}T_0' + s^{m'+1}T_1' + s^{m'+2}T_2' + \cdots$$

and

$$S''' = s^{m'''}T_0''' + s^{m'''+1}T_1''' + s^{m'''+2}T_2''' \cdots$$

are two expressions of the form S, and let us suppose, further, that the coefficient r_0' of T_0' is different from zero. By equating the like powers of s in the two members of the equation

$$S'S'' = S''',$$

we find, first of all, in a definite manner an integral number m'' as exponent, and then such a succession of expressions

$$T_0'', \ T_1'', \ T_2'' \dots$$

that, by aid of formula (1), the expression

$$S'' = s^{m''} T_0'' + s^{m''+1} T_1'' + s^{m''+2} T_2'' \dots$$

satisfies equation (2). With this our theorem is established.

In order, finally, to render possible an order of sequence of the numbers of our system $\Omega(s,t)$, we make the following conventions. Let a number of this system be called greater or less than 0 according as in the expression S, which represents it, the first coefficient r_0 of T_0 is greater or less than zero. Given any two numbers a, b of the complex number system under consideration, we say that $a < b$ or $a > b$ according as we have $a - b < 0$ or > 0. It is seen immediately that, with these conventions, the laws 13–16 of § 13 are valid; that is to say, $\Omega(s,t)$ is a desarguesian number system (see § 28).

As equation (1) shows, law 12 of § 13 is not fulfilled by our complex number system and, consequently, the validity of theorem 39 is fully established.

In conformity with theorem 38, Archimedes's theorem (theorem 17, §13) does not hold for the number system $\Omega(s,t)$ which we have just constructed.

We wish also to call attention to the fact that the number system $\Omega(s,t)$, as well as the systems O and $\Omega(t)$ made use of in § 9 and § 12, respectively, contains only an enumerable set of numbers.

§34. PROOF OF THE TWO PROPOSITIONS CONCERNING PASCAL'S THEOREM (NON-PASCALIAN GEOMETRY)

If, in a geometry of space, all of the axioms I, II, III are fulfilled, then Desargues's theorem (theorem 32) is also valid, and, consequently, according to §§ 24–26, pp. 50–58, it is possible to introduce into this geometry an algebra of segments for which the rules 1–11, 13–16 of § 13 are all valid. If we assume now that the axiom (V) of Archimedes is valid for our geometry, then evidently Archimedes's theorem (theorem 17 of § 13) also holds for our algebra of segments, and, consequently, by virtue of theorem 38, the commutative law of multiplication is valid. Since, however, the definition of the product of two segments, as introduced in § 24 (figure 42) and which is the definition here also under discussion, agrees with the definition in § 15 (figure 22), it follows from the construction made in § 15 that the commutative law of multiplication is here nothing else than Pascal's theorem. Consequently, the validity of theorem 36 is established.

In order to demonstrate theorem 37, let us consider again the desarguesian number system $\Omega(s, t)$ introduced in §33, and construct, in the manner described in § 29, a geometry of space for which all of the axioms I, II, III are fulfilled. However, Pascal's theorem will not hold for this geometry; for, the commutative law of multiplication is not valid in the desarguesian number system $\Omega(s, t)$. According to theorem 36, the non-pascalian geometry is then necessarily also a non-archimedean geometry.

By adopting the hypothesis we have, it is evident that we cannot demonstrate Pascal's theorem, providing we regard our geometry of space as a part of a geometry of an arbitrary number of dimensions in which, besides the points, straight lines, and planes, still other linear elements are present, and providing there exists for these elements a corresponding system of axioms of connection and of order, as well as the axiom of parallels.

§35. THE DEMONSTRATION, BY MEANS OF THE THEOREMS OF PASCAL AND DESARGUES, OF ANY THEOREM RELATING TO POINTS OF INTERSECTION

Every proposition relating to points of intersection in a plane has necessarily The form: Select, first of all, an arbitrary system of points and straight lines satisfying respectively the condition that certain ones of these points are situated on certain ones of the straight lines. If, in some known manner, we construct the straight lines joining the given points and determine the points of intersection of the given lines, we shall obtain finally a definite system of three straight lines, of which our proposition asserts that they all pass through the same point.

Suppose we now have a plane geometry in which all of the axioms I 1–2, II . . . , V are valid. According to § 17, pp. 33–35, we may now find, by making use of a rectangular pair of axes, for each point a corresponding pair of numbers (x, y) and for each straight line a ratio of three definite numbers $(u : v : w)$. Here, the numbers x, y, u, v, w are all *real* numbers, of which u, v cannot both be zero. The condition showing that the given point is situated upon the given straight line, viz.:

$$ux + vy + w = 0 \qquad (14)$$

is an equation in the ordinary sense of the word. Conversely, in case x, y, u, v, w are numbers of the algebraic domain Ω of § 9, and u, v are not both zero, we may certainly assume that each pair of numbers (x, y) gives a point and that each ratio of three numbers $(u : v : w)$ gives a straight line in the geometry in question.

If, for all the points and straight lines which occur in connection with any theorem relating to intersections in a plane, we introduce the corresponding pairs and triples of numbers, then such a theorem asserts that a definite expression $A(p_1, p_2, ..., p_r)$ with real coefficients and depending rationally upon certain parameters $p_1, p_2, ..., p_r$ always vanishes as soon as we put for each of these parameters a number of the main Ω considered in § 9. We conclude from this that the expression $A(p_1, p_2, ..., p_r)$ must also vanish identically in accordance with the laws 7–12 of § 13.

Since, according to § 32, Desargues's theorem holds for the geometry in question, it follows that we certainly can make use of the algebra of segments introduced in §24, and because Pascal's theorem is equally valid in this case, the commutative law of multiplication is also. Hence, for this algebra of segments, all of the laws 7–12 of § 13 are valid.

If we take as our axes in this new algebra of segments the co-ordinate axes already used and consider the unit points E, E' as suitably established, we see that the new algebra of segments is nothing else than the system of co-ordinates previously employed.

In order to show that, for the new algebra of segments, the expression $A(p_1, p_2, ..., p_r)$ vanishes identically, it is sufficient to apply the theorems of Pascal and Desargues. Consequently we see that:

Every proposition relative to points of intersection in the geometry in question must always, by the aid of suitably constructed auxiliary points and straight lines, turn out to be a combination of the theorems of Pascal and Desargues. Hence for the proof of the validity of a theorem relating to points of intersection, we need not have resource to the theorems of congruence.

7

GEOMETRICAL CONSTRUCTIONS BASED UPON THE AXIOMS I–V

§36. GEOMETRICAL CONSTRUCTIONS BY MEANS OF A STRAIGHT-EDGE AND A TRANSFERER OF SEGMENTS

Suppose we have given a geometry of space, in which all of the axioms I–V are valid. For the sake of simplicity, we shall consider in this chapter a a plane geometry which is contained in this geometry of space and shall investigate the question as to what elementary geometrical constructions may be carried out in such a geometry.

Upon the basis of the axioms of group I, the following constructions are always possible.

Problem 1

To join two points with a straight line and to find the intersection of two straight lines, the lines not being parallel.

Axiom III renders possible the following construction:

Problem 2

Through a given point to draw a parallel to a given straight line.

By the assistance of the axioms (IV) of congruence, it is possible to lay off segments and angles; that is to say, in the given geometry we may solve the following problems:

Problem 3

To lay off from a given point upon a given straight line a given segment.

Problem 4

To lay off on a given straight line a given angle; or what is the same thing, to construct a straight line which shall cut a given straight line at a given angle.

It is impossible to make any new constructions by the addition of the axioms of groups II and V. Consequently, when we take into consideration merely the axioms of groups I–V, all of those constructions and only those are possible, which may be reduced to the problems 1–4 given above.

We will add to the fundamental problems 1–4 also the following:

Problem 5

To draw a perpendicular to a given straight line.

We see at once that this construction can be made in different ways by means of the problems 1–4.

In order to carry out the construction in problem 1, we need to make use of only a *straight edge*. An instrument which enables us to make the construction in problem 3, we will call a *transferer of segments*. We shall now show that problems 2, 4, and 5 can be reduced to the constructions in problems 1 and 3 and, consequently, all of the problems 1–5 can be completely constructed by means of a straight-edge and a transferer of segments.

We arrive, then, at the following result:

Theorem 40

Those problems in geometrical construction, which may be solved by the assistance of only the axioms I–V, can always be carried out by the use of the straight-edge and the transferer of segments.

Proof In order to reduce problem 2 to the solution of problems 1 and 3, we join the given point P with any point A of the given straight line and produce PA to C, making $AC = PA$. Then, join C with any other point B of the given straight line and produce CB to Q, making, $BQ = CB$. The straight line PQ is the desired parallel.

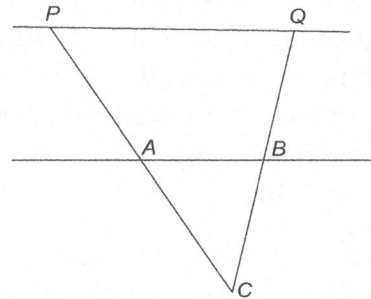

Figure 50

We can solve problem 5 in the following manner. Let A be an arbitrary point of the given straight line. Then upon this straight line, lay off in both directions from *A* the two equal segments *AB* and *AC*. Determine, upon any two straight lines passing through the point *A*, the points *E* and *D* so that the segments *AD* and *AE* will equal *AB* and *AC*.

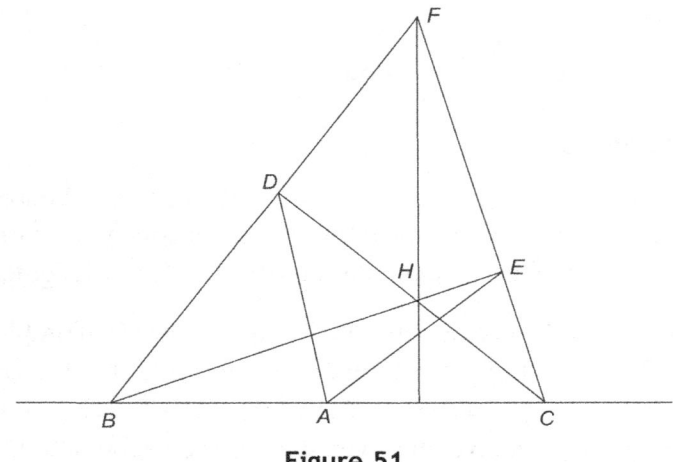

Figure 51

Suppose the straight lines *BD* and *CE* intersect in *F* and the straight lines *BE* and *CD* intersect in *H*. *FH* is then the desired perpendicular. In fact, the angles *BDC* and *BEC*, being inscribed in a semicircle having the diameter *BC*, are both right angles, and, hence, according to the theorem relating to the point of intersection of the altitudes of a triangle, the straight lines *FH* and *BC* are perpendicular to each other.

Moreover, we can easily solve problem 4 simply by the drawing of straight lines and the laying off of segments. We will employ the following method which requires only the drawing of parallel lines and the erection of perpendiculars. Let *β* be the angle to be laid off and *A* its vertex. Draw through *A* a straight line *l* parallel to the given straight line, upon which we are to lay off the given angle β. From an arbitrary point B of one side of the angle β, let fall a perpendicular upon the other side of this angle and also one upon l.

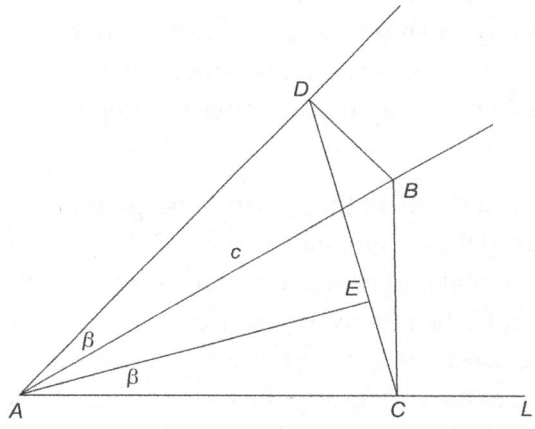

Figure 52

Denote the feet of these perpendiculars by D and C respectively. The construction of these perpendiculars is accomplished by means of problems 2 and 5. Then, let fall from A a perpendicular upon CD, and let its foot be denoted by E. According to the demonstration given in Section § 14, the angle CAE equals β. Consequently, the construction in 4 is made to depend upon that of 1 and 3 and with this our proposition is demonstrated.

§37. ANALYTICAL REPRESENTATION OF THE CO-ORDINATES OF POINTS WHICH CAN BE SO CONSTRUCTED

Besides the elementary geometrical problems considered in § 36, there exists a long series of other problems whose solution is possible by the drawing of straight lines and the laying off of segments.

In order to get a general survey of the scope of the problems which may be solved in this manner, let us take as the basis of our consideration a system of axes in rectangular co-ordinates and suppose that the co-ordinates of the points are, as usual, represented by real numbers or by functions of certain arbitrary parameters. In order to answer the question in respect to all the points capable of such a construction, we employ the following considerations.

Let a system of definite points be given. Combine the co-ordinates of these points into a domain R. This domain contains, then, certain real numbers and certain arbitrary parameters p. Consider, now, the totality of points capable of construction by the drawing of straight lines and the laying off of definite segments, making use of the system of points in question. We will call the domain formed from the co-ordinates of these points $\Omega(R)$, which will then contain real numbers and functions of the arbitrary parameters p.

The discussion in § 17 shows that the drawing of straight lines and of parallels amounts, analytically, to the addition, subtraction, multiplication, and division of segments. Furthermore, the well known formula given in § 9 for a rotation shows that the laying off of segments upon a straight line does not necessitate any other analytical operation than the extraction of the square root of the sum of the squares of two segments whose bases have been previously constructed. Conversely, in consequence of the pythagorean theorem, we can always construct, by the aid of a right triangle, the square root of the sum of the squares of two segments by the mere laying off of segments.

From these considerations, it follows that the domain $\Omega(R)$ contains all of those and only those real numbers and functions of the parameters p, which arise from the numbers and parameters in R by means of a finite number of applications of the five operations; viz., the four elementary operations of arithmetic and, in addition, the fifth operation of extracting the square root of the sum of two squares. We may express this result as follows:

Theorem 41

A problem in geometrical construction is, then, possible of solution by the drawing of straight lines and the laying off of segments, that is to say, by the use of the straight-edge and a transferer of segments, when and only when, by the analytical solution of the problem, the co-ordinates of the desired points are such functions of the co-ordinates of the given points as may be determined by the rational operations and, in addition, the extraction of the square root of the sum of two squares.

From this proposition, we can at once show that not every problem which can be solved by the use of a compass can also be solved by the aid of a transferer of segments and a straight-edge. For the purpose of showing this, let us consider again that geometry which was constructed in § 9 by the help of the domain Ω of algebraic numbers. In this geometry, there exist only such segments as can be constructed by means of a straight-edge and a transferer of segments, namely, the segments determined by the numbers of the domain Ω.

Now, if ω is a number of the domain Ω, we easily see from the definition of Ω that every algebraic number conjugate to ω must also lie in Ω. Since the numbers of the domain Ω are evidently all real, it follows that it can contain only such real algebraic numbers as have their conjugates also real.

Let us now consider the following problem; viz., to construct a right triangle having the hypotenuse 1 and one side $\left| \sqrt{2} \right| - 1$. The algebraic number $\sqrt{2 \left| \sqrt{2} \right| - 2}$, which expresses the numerical value of the other side, does not occur in the domain Ω, since the conjugate number $\sqrt{-2 \left| \sqrt{2} \right| - 2}$ is imaginary. This problem is, therefore, not capable of solution in the geometry in question and, hence, cannot be constructed by means of a straight-edge and a transferer of segments, although the solution by means of a compass is possible.

§38. THE REPRESENTATION OF ALGEBRAIC NUMBERS AND OF INTEGRAL RATIONAL FUNCTIONS AS SUMS OF SQUARES

The question of the possibility of geometrical constructions by the aid of a straight-edge and a transferer of segments necessitates, for its complete treatment, particular theorems of an arithmetical and algebraic character, which, it appears to me, are themselves of interest. Since the time of Fermat, it has been known that every positive integral rational number can be represented as the sum of four squares. This theorem of Fermat permits the following remarkable generalization:

Definition

Let k be an arbitrary number field and let m be its degree. We will denote by k', k'', ..., $k^{(m-1)}$ the $m - 1$ number fields conjugate to k. If, among the m fields $k, k', k'', ..., k^{(m-1)}$ there is one or more formed entirely of real numbers, then we call these fields real. Suppose that the fields k, k', ..., $k^{(s-1)}$ are such. A number α of the field k is called in this case *totally positive in k,* whenever the s numbers conjugate to α, contained respectively in k, k', k'', ..., $k^{(s-1)}$ are all positive. However, if in each of the m fields k, k', k'', ..., $k^{(m-1)}$ there are also imaginary numbers present, we call every number α in k *totally positive*.

We have, then, the following proposition:

Theorem 42

Every totally positive number in k may be represented as the sum of four squares, whose bases are integral or fractional numbers of the field k.

The demonstration of this theorem presents serious difficulty. It depends essentially upon the theory of relatively quadratic number fields, which I have recently developed in several papers.[17] We will here call attention only to that proposition in this theory which gives the condition that a ternary diophantine equation of the form

$$\alpha\xi^2 + \beta\eta^2 + \gamma\zeta^2 = 0$$

can be solved when the coefficients α, β, γ are given numbers in k and ξ, η, ζ are the required numbers in k. The demonstration of theorem 42 is accomplished by the repeated application of the proposition just mentioned.

From theorem 42 follow a series of propositions concerning the representation of such rational functions of a variable, with rational coefficients, as never have negative values. I will mention only the following theorem, which will be of service in the following sections.

Theorem 43

Let, $f(x)$ be an integral rational function of x whose coefficients are rational numbers and which never becomes negative for any real value of x. Then $f(x)$ can always be represented as the quotient of two sums of squares of which the bases are all integral rational functions of x with rational coefficients.

Proof   We will denote the degree of the function $f(x)$ by m, which, in any case, must evidently be even. When $m = 0$, that is to say, when $f(x)$ is a rational number, the validity of theorem 43 follows immediately from Fermat's theorem concerning the representation of a positive number as the sum of four squares. We will assume that the proposition is already established for functions of degree 2, 4, 6, . . . , $m - 2$, and show, in the following manner, its validity for the case of a function of the mth degree.

Let us, first of all, consider briefly the case where $f(x)$ breaks up into the product of two or more integral functions of x with rational coefficients. Suppose $p(x)$ to be one of those functions contained in $f(x)$, which itself cannot be further decomposed into a product of integral functions having rational coefficients. It then follows at once from the "definite" character which we have given to the function $f(x)$, that the factor $p(x)$ must either appear in $f(x)$ to an even degree or $p(x)$ must be itself "definite"; that is to say, must be a function which never has negative values for any real values of x. In the first case, the quotient $\dfrac{f(x)}{\left\{p(x)^2\right\}}$ and, in the second case, both $p(x)$ and $\dfrac{f(x)}{p(x)}$, are "definite," and these functions have an even degree $< m$. Hence, according to our hypothesis, in the first case, $\dfrac{f(x)}{\left\{p(x)\right\}^2}$ and, in the last case, $p(x)$

and $\dfrac{f(x)}{p(x)}$ may be represented as the quotient of the sum of squares of the character mentioned in theorem 43. Consequently, in both of these cases, the function $f(x)$ admits of the required representation.

Let us now consider the case where $f(x)$ cannot be broken up into the product of two integral functions having rational coefficients. The equation $f(\theta) = 0$ defines, then, a field of algebraic numbers $k(\theta)$ of the mth degree, which, together with all their conjugate fields, are imaginary. Since, according to the definition given just before the statement of theorem 42, each number given in $k(\theta)$, and hence also -1 is totally positive in $k(\theta)$, it follows from theorem 42 that the number -1 can be represented as a sum of the squares of four definite numbers in $k(\theta)$. Let, for example

$$-1 = \alpha^2 + \beta^2 + \gamma^2 + \delta^2, \tag{1}$$

where $\alpha, \beta, \gamma, \delta$ are integral or fractional numbers in $k(\theta)$. Let us put

$$\alpha = a_1\theta^{m-1} + a_2{}^{m-2} + \cdots + a_m = \phi(\theta),$$
$$\beta = b_1\theta^{m-1} + b_2\theta^{m-2} + \cdots + b_m = \psi(\theta),$$
$$\gamma = c_1\theta^{m-1} + c_2\theta^{m-2} + \cdots + c_m = \chi(\theta),$$
$$\theta = d_1\theta^{m-1} + d_2\theta^{m-2} + \cdots + d_m = \rho(\theta);$$

where $a_1, a_2, ..., a_m, ..., d_1, d_2, ..., d_m$ are the rational numerical coefficients and $\phi(\theta), \psi(\theta), \chi(\theta), \rho(\theta)$ the integral rational functions in question, having the degree $(m-1)$ in θ.

From (1), we have

$$1 + \{\phi(\theta)\}^2 + \{\psi(\theta)\}^2 + \{\chi(\theta)\}^2 + \{\rho(\theta)\}^2 = 0$$

Because of the irreducibility of the equation $f(x) = 0$, the expression

$$F(x) = 1 + \{\phi(\theta)\}^2 + \{\psi(\theta)\}^2 + \{\chi(\theta)\}^2 + \{\rho(\theta)\}^2$$

represents, necessarily, an integral rational function of x which is divisible by $f(x) \cdot F(x)$ is, then, a "definite" function of the degree $(2m - 2)$ or lower. Hence, the quotient $\dfrac{F(x)}{f(x)}$ is a "definite" function of the degree $(m - 2)$ or lower in x, having rational coefficients. Consequently, by the hypothesis we have made, $\dfrac{F(x)}{f(x)}$ can be represented as the quotient of two sums of squares of the kind mentioned in theorem 43 and, since $F(x)$ is itself such a sum of squares, it follows that, $f(x)$ must also be a quotient of two sums of squares of the required kind. The validity of theorem 43 is accordingly established.

It would be perhaps difficult to formulate and to demonstrate the corresponding proposition for integral functions of two or more variables. However, I will here merely remark that I have demonstrated in an entirely different manner the possibility of representing any "definite" integral rational function of two variables as the quotient of sums of squares of integral functions, upon the hypothesis that the functions represented may have as coefficients not only rational but *any* real numbers.[18]

§39. CRITERION FOR THE POSSIBILITY OF A GEOMETRICAL CONSTRUCTION BY MEANS OF A STRAIGHT-EDGE AND A TRANSFERER OF SEGMENTS

Suppose we have given a problem in geometrical construction which can be affected by means of a compass. We shall attempt to find a criterion which will enable us to decide, from the analytical nature of the problem and its solutions, whether or not the construction can be carried out by means of only a straight-edge and a transferer of segments. Our investigation will lead us to the following proposition.

Theorem 44

Suppose we have given a problem in geometrical construction, which is of such a character that the analytical treatment of it enables us to determine uniquely the co-ordinates of the desired points from the co-ordinates of the given points by means of the rational operations and the extraction of the square root. Let n be the smallest number of square roots which suffice to calculate the co-ordinates of the points. Then, in order that the required construction shall be possible by the drawing of straight lines and the laying off of segments, it is necessary and sufficient that the given geometrical problem shall have exactly 2^n real solutions for every position of the given points; that is to say, for all values of the arbitrary parameter expressed in terms of the co-ordinates of the given points.

Proof   We shall demonstrate this proposition merely for the case where the coordinates of the given points are rational functions, having rational coefficients, of a single parameter p.

It is at once evident that the proposition gives a necessary condition. In order to show that it is also sufficient, let us assume that it is fulfilled and then, among the n square roots, consider that one which, in the calculation of the co-ordinates of the desired points, is first to be extracted. The expression under this radical is a rational function $f_1(p)$, having rational coefficients, of the parameter p. This rational function cannot have a negative value for any real value of the parameter p; for, otherwise the problem must have imaginary solutions for certain values of p, which is contrary to the given hypothesis. Hence, from theorem 43, we conclude that $f_1(p)$ can be represented as a quotient of the sums of squares of integral rational functions.

Moreover, the formulæ

$$\sqrt{a^2 + b^2 + c^2} = \sqrt{\left(\sqrt{a^2 + b^2}\right)^2 + c^2}$$

$$\sqrt{a^2 + b^2 + c^2 + d^2} = \sqrt{\left(\sqrt{a^2 + b^2 + c^2}\right)^2 + d^2}$$

show that, in general, the extraction of the square root of a sum of any number of squares may always be reduced to the repeated extraction of the square root of the sum of two squares.

If now we combine this conclusion with the preceding results, it follows that the expression $\sqrt{f_1(p)}$ can certainly be constructed by means of a straight-edge and a transferer of segments. Among the n square roots, consider now the second one to be extracted in the process of calculating the co-ordinates of the required points. The expression under this radical is a rational function $f_2\left(p, \sqrt{f_1}\right)$ of the parameter p and the square root first considered. This function f_2 can never be negative for any real arbitrary value of the parameter p and for either sign of $\sqrt{f_1}$; for, otherwise among the 2^n solutions of our problem, there would exist for certain values of p also imaginary solutions, which is contrary to our hypothesis. It follows, therefore, that f_2 must satisfy a quadratic equation of the form

$$f_2^2 - \phi_2(p)f_2 + \psi_1(p) = 0,$$

where $\phi_1(p)$ and $\psi_1(p)$ are, necessarily, such rational functions of p as have rational coefficients and for real values of p never become negative. From this equation, we have

$$f_2 = \frac{f_2^2 + \psi_1(p)}{\phi_1(p)}$$

Now, according to theorem 43, the functions $\phi_1(p)$ and $\psi_1(p)$ must again be the quotient of the sums of squares of rational functions, and, on the other hand, the expression f_2 may be, from the above considerations, constructed by means of a straight-edge and a transferer of segments. The expression found for f_2 shows, therefore, that f_2 is a quotient of the sum of squares of functions which may be constructed in the same way. Hence, the expression $\sqrt{f_2}$ can also be constructed by means of a straight-edge and a transferer of segments.

Just as with the expression f_2, any other rational function $\phi_2\left(p, \sqrt{f_1}\right)$ of p and $\sqrt{f_1}$ may be shown to be the quotient of two sums of squares of functions which may be constructed, provided this rational function ϕ_2 possesses the property that for real values of the parameter p and for either sign of $\sqrt{f_1}$, it never becomes negative.

This remark permits us to extend the above method of reasoning in the following manner.

Let $f_3\left(p, \sqrt{f_1}, \sqrt{f_2}\right)$ be such an expression as depends in a rational manner upon the three arguments p, $\sqrt{f_1}$, $\sqrt{f_2}$ and of which, in the analytical calculation of the co-ordinates of the desired points, the square root is the third to be extracted. As before, it follows that f_3 can never have negative values for real values of p and for either sign of $\sqrt{f_1}$ and $\sqrt{f_2}$. This condition of affairs shows again that f_3 must satisfy a quadratic equation of the form

$$f_3^2 - \phi_2\left(p, \sqrt{f_1}\right)f_3 - \psi_2\left(p, \sqrt{f_1}\right) = 0,$$

where ϕ_2 and ψ_2 are such rational functions of p and $\sqrt{f_1}$ as never become negative for any real value of p and either sign of $\sqrt{f_1}$. But, according to the preceding remark, the functions ϕ_2 and ψ_2 are the quotients of two sums of squares of functions which may be constructed and, hence, it follows that the expression

$$f_3 = \frac{f_3^2 + \psi_2(p,\sqrt{f_1})}{\phi_2(p,\sqrt{f_1})}$$

is likewise possible of construction by aid of a straight-edge and a transferer of segments.

The continuation of this method of reasoning leads to the demonstration of theorem 44 for the case of a single parameter p.

The truth of theorem 44 for the general case depends upon whether or not theorem 43 can be generalized in a similar manner to cover the case of two or more variables.

As an example of the application of theorem 44, we may consider the regular polygons which may be constructed by means of a compass. In this case, the arbitrary parameter, p does not occur, and the expressions to be constructed all represent algebraic numbers. We easily see that the criterion of theorem 44 is fulfilled, and, consequently, it follows that the above-mentioned regular polygons can be constructed by the drawing of straight lines and the laying off of segments. We might deduce this result also directly from the theory of the division of the circle (*Kreisteilung*).

Concerning the other known problems of construction in the elementary geometry, we will here only mention that the problem of Malfatti may be constructed by means of a straight-edge and a transferer of segments. This is, however, not the case with the contact problems of Appolonius.

CONCLUSION

The preceding work treats essentially of the problems of the euclidean geometry only; that is to say, it is a discussion of the questions which present themselves when we admit the validity of the axiom of parallels. It is none the less important to discuss the principles and the fundamental theorems when we disregard the axiom of parallels. We have thus excluded from our study the important question as to whether it is possible to construct a geometry in a logical manner, without introducing the notion of the plane and the straight line, by means of only points as elements, making use of the idea of groups of transformations, or employing the idea of distance. This last question has recently been the subject of considerable study, due to the fundamental and prolific works of Sophus Lie. However, for the complete elucidation of this question, it would be well to divide into several parts the axiom of Lie, that space is a numerical multiplicity. First of all, it would seem to me desirable to discuss thoroughly the hypothesis of Lie, that functions which produce transformations are not only continuous, but may also be differentiated.

As to myself, it does not seem to me probable that the geometrical axioms included in the condition for the possibility of differentiation are all necessary. In the treatment of all questions of this character, I believe the methods and the principles employed in the preceding work will be of value. As an example, let me call attention to an investigation undertaken at my suggestion by Mr. Dehn, and which has already appeared.[19] In this article, he has discussed the known theorems of Legendre concerning the sum of the angles of a triangle, in the demonstration of which that geometer made use of the idea of continuity.

The investigation of Mr. Dehn rests upon the axioms of connection, of order, and of congruence; that is to say, upon the axioms of groups I, II, IV. However, the axiom of parallels and the axiom of Archimedes are excluded. Moreover, the axioms of order are stated in a more general manner than in the present work, and in substance as follows: Among four points A, B, C, D of a straight line, there are always two, for example A, C, which are separated from the other two and conversely. Five points A, B, C, D, E upon a straight line may always be so arranged that A, C shall be separated from B, E and from B, D. Consequently, A, D are always separated from B, E and from C, E, etc. The (elliptic) geometry of Riemann, which we have not considered in the present work, is in this way not necessarily excluded.

Upon the basis of the axioms of connection, order, and congruence, that is to say, the axioms I, II, IV, we may introduce, in the well known manner, the elements called ideal,—-ideal points, ideal straight lines, and ideal planes. Having done this, Mr. Dehn demonstrates the following theorem.

If, with the exception of the straight line t and the points lying upon it, we regard all of the straight lines and all of the points (ideal or real) of a plane as the elements of a new geometry, we may then define a new kind of congruence so that all of the axioms of connection, order, and congruence, as well as the axiom of Euclid, shall be fulfilled. In this new geometry, the straight line t takes the place of the straight line at infinity.

This euclidean geometry, superimposed upon the non-euclidean plane, may be called a *pseudo-geometry* and the new kind of congruence a *pseudo-congruence*.

By aid of the preceding theorem, we may now introduce an algebra of segments relating to the plane and depending upon the developments made in §15, pp. 29–31. This algebra of segments permits the demonstration of the following important theorem:

If, in any triangle whatever, the sum of the angles is greater than, equal to, or less than, two right angles, then the same is true for all triangles.

The case where the sum of the angles is equal to two right angles gives the well known theorem of Legendre. However, in his demonstration, Legendre makes use of continuity.

Mr. Dehn then discusses the connection between the three different hypotheses relative to the sum of the angles and the three hypotheses relative to parallels.

He arrives in this manner at the following remarkable propositions.

Upon the hypothesis that through a given point we may draw an infinity of lines parallel to a given straight line, it does not follow, when we exclude the axiom of Archimedes, that the sum of the angles of a triangle is less than two right angles, but on the contrary, this sum may be

(a) greater than two right angles, or

(b) equal to two right angles.

In order to demonstrate part (a) of this theorem, Mr. Dehn constructs a geometry where we may draw through a point an infinity of lines parallel to a given straight line and where, moreover, all of the theorems of Riemann's (elliptic) geometry are valid. This geometry may be called non-legendrian, for it is in contradiction with that theorem of Legendre by virtue of which the sum of the angles a triangle is never greater than two right angles. From the existence of this non-legendrian geometry, it follows at once that it is impossible to demonstrate the theorem of Legendre just mentioned without employing the axiom of Archimedes, and in fact, Legendre made use of continuity in his demonstration of this theorem.

For the demonstration of case (b), Mr. Dehn constructs a geometry where the axiom of parallels does not hold, but where, nevertheless, all of the theorems of the euclidean geometry are valid. Then, we have the sum of the angles of a triangle equal to two right angles. There exist also similar triangles, and the extremities of the perpendiculars having the same length and their bases upon a straight line all lie upon the same straight line, etc. The existence of this geometry shows that, if we disregard the axiom of Archimedes, the axiom of parallels cannot be replaced by any of the propositions which we usually regard as equivalent to it.

This new geometry may be called a *semi-euclidean geometry*. As in the case of the non-legendrian geometry, it is clear that the semi-euclidean geometry is at the same time a non-archimedean geometry.

Mr. Dehn finally arrives at the following surprising theorem:

Upon the hypothesis that there exists no parallel, it follows that the sum of the angles of a triangle is greater than two right angles.

This theorem shows that, with respect to the axiom of Archimedes, the two non-euclidean hypotheses concerning parallels act very differently.

We may combine the preceding results in the following table.

The sum of the angles of a triangle is	Through a given point, we may draw		
	No parallels to a straight line	One parallel to a straight line	An infinity of parallels to a straight line
> 2 right angles	Riemann's (elliptic) geometry	This case is impossible	Non-legendrian geometry
< 2 right angles	This case is impossible	Euclidean (parabolic) geometry	Semi-euclidean geometry
= 2 right angles	This case is impossible	This case is impossible	Geometry of Lobatschewski (hyperbolic)

However, as I have already remarked, the present work is rather a critical investigation of the principles of the euclidean geometry. In this investigation, we have taken as a guide the following fundamental principle; viz., to make the discussion of each question of such a character as to examine at the same time whether or not it is possible to answer this question by following out a previously determined method and by employing certain limited means. This fundamental rule seems to me to contain a general law and to conform to the nature of things. In fact, whenever in our mathematical investigations we encounter a problem or suspect the existence of a theorem, our reason is satisfied only when we possess a complete solution of the problem or a rigorous demonstration of the theorem, or, indeed, when we see clearly the reason of the impossibility of the success and, consequently, the necessity of failure.

Thus, in the modern mathematics, the question of the impossibility of solution of certain problems plays an important role, and the attempts made to answer such questions have often been the occasion of discovering new and fruitful fields for research. We recall in this connection the demonstration by Abel of the impossibility of solving an equation of the fifth degree by means of radicals, as also the discovery of the impossibility of demonstrating the axiom of parallels, and, finally, the theorems of Hermite and Lindeman concerning the impossibility of constructing by algebraic means the numbers e and π.

This fundamental principle, which we ought to bear in mind when we come to discuss the principles underlying the impossibility of demonstrations, is intimately connected with the condition for the "purity" of methods in demonstration, which in recent times has been considered of the highest importance by many mathematicians. The foundation of this condition is nothing else than a subjective conception of the fundamental principle given above. In fact, the preceding geometrical study attempts, in general, to explain what are the axioms, hypotheses, or means, necessary to the demonstration of a truth of elementary geometry, and it only remains now for us to judge from the point of view in which we place ourselves as to what are the methods of demonstration which we should prefer.

APPENDIX[20]

The investigations by Riemann and Helmholtz of the foundations of geometry led Lie to take up the problem of the *axiomatic* treatment of geometry as introductory to the study of groups. This profound mathematician introduced a system of axioms which he showed by means of his theory of transformation groups to be sufficient for the complete development of geometry.[21]

As the basis of his transformation groups, Lie made the assumption that the functions defining the group can be differentiated. Hence in Lie's development, the question remains uninvestigated as to whether this assumption as to the differentiability of the functions in question is really unavoidable in developing the subject according to the axioms of geometry, or whether, on the other hand, it is not a consequence of the group-conception and of the remaining axioms of geometry. In consequence of his method of development, Lie has also necessitated the express statement of the axiom that the group of displacements is produced by infinitesimal transformations. These requirements, as well as essential parts of Lie's fundamental axioms concerning the nature of the equation defining points of equal distance, can be expressed geometrically in only a very unnatural and complicated manner. Moreover, they appear only through the analytical method used by Lie and not as a necessity of the problem itself.

In what follows, I have therefore attempted to set up for plane geometry a system of axioms, depending likewise upon the conception of a group,[22] which contains only those requirements which are simple and easily seen geometrically. In particular they do not require the differentiability of the functions defining displacement. The axioms of the system which I set up are a special division of Lie's, or, as I believe, are at once deducible from his.

My method of proof is entirely different from Lie's method. I make use particularly of Cantor's theory of assemblages of points and of the theorem of C. Jordan, according to which every closed continuous plane curve free from double points divides the plane into an inner and an outer region.

To be sure, in the system set up by me, particular parts are unnecessary. However, I have turned aside from the further investigation of these conditions to the simple statement of the axioms, and above all because I wish to avoid a comparatively too complicated proof, and one which is not at once geometrically evident.

In what follows I shall consider only the axioms relating to the plane, although I suppose that an analogous system of axioms for space can be set up which will make possible the construction of the geometry of space in a similar manner.

We establish the following convention, namely: We will understand by *number-plane* the ordinary plane having a rectangular system of co-ordinates x, y.

A continuous curve lying in this number-plane and being free from double points and including its end points is called a *Jordan curve*. If the Jordan curve is closed, the interior of the region of the number-plane bounded by it is called a *Jordan region*.

For the sake of easier representation and comprehension, I shall in the following investigation formulate the definition of the plane in a more restricted sense than my method of proof requires,[23] namely: I shall assume that it is possible to map[24] in a reversible, single-valued manner all of the points of our geometry at the same time upon the points lying in the finite region of the number-plane, or upon a definite partial system of the same. Hence, each point of our geometry is characterized by a definite pair of numbers x, y. We formulate this statement of the idea of the plane as follows:

Definitions of the Plane

The plane is a system of points which can be mapped in a reversible, single-valued manner upon the points lying in the finite region of the number-plane, or upon a certain partial system of the same. To each point A of our geometry, there exists a Jordan curve in whose interior the map of A lies and all of whose points likewise represent points of our geometry. This Jordan region is called the domain of the point A. Each Jordan region contained in a Jordan region which includes the point A is likewise called a domain of A. If B is any point in a domain of A, then this domain is at the same time called also a domain of B.

If A and B are any two points of our geometry, then there always exists a domain which contains at the same time both of the points A and B.

We will define a *displacement* as a reversible, single-valued transformation of a plane into itself. Evidently we may distinguish two kinds of reversible, single-valued, continuous transformations of the number-plane into itself. If we take any closed Jordan curve in the number-plane and think of its being traversed in a definite sense, then by such a transformation this curve goes over into another closed Jordan curve which is also traversed in a certain sense. We shall assume in the present investigation that it is traversed in the same sense as the original Jordan curve, when we apply a transformation of the number-plane into itself, which defines a displacement. This assumption[25] necessitates the following statement of the conception of a displacement.

Definition of Displacement

A displacement is a reversible, single-valued, continuous transformation of the maps of the given points upon the number-plane into themselves in such a manner that a closed Jordan curve is traversed in the same sense after the transformation as before. A displacement by which the point M remains unchanged is called a *rotation*[26] about the point M.

In accordance with the conventions setting forth the notions "plane" and "displacement," we set up the three following axioms:

Axiom I   *If two displacements are followed out one after the other, then the resulting map of the plane upon itself is again a displacement.*

We say briefly:

Axiom I   The displacements form a group.

Axiom II   *If A and M are two arbitrary points distinct from each other, then by a rotation about M we can always bring A into an infinite number of different positions.*

If in our geometry we define a true circle as the totality of those points which arise by rotating about M a point different from M, then we can express the statement made in axiom II as follows:

Axiom II   Every true circle consists of an infinite number of points.

As preliminary to axiom III, we make the following explanations:

Let A be a definite point in our geometry and $A_1, A_2, A_3, \ldots$ any infinite system of points. With the same letters we will also denote the maps of these points upon the number-plane. About the point A in the number-plane take an arbitrarily small domain α. If then any of the map-points A_i fall within the domain α, we say that there are points A_i arbitrarily near the point A.

Let A, B be a definite pair of points in our geometry, and let $A_1 B_1, A_2 B_2, A_3 B_3, \ldots$ be any infinite system of pairs of points. With the same letters we will denote the maps of these pairs of points upon the number-plane. Select about each of the points A and B in the number-plane an arbitrarily small domain α and β, respectively. If then there are pairs of points $A_i B_i$ such that A_i falls within the domain α and at the same time B_i falls within the domain β, we say that there are segments $A_i B_i$ lying arbitrarily near the segment AB.

Let ABC a definite triad of points in our geometry, and let $A_1 B_1 C_1, A_2 B_2 C_2, A_3 B_3 C_3, \ldots$ be any infinite system of triads of points. With the same letters we will also denote the maps of these triads of points upon the number-plane. About each of the points A, B, C in the number-plane take an arbitrarily small domain α, β, γ respectively. If then there are triads of points $A_i B_i C_i$ such that A_i falls in the domain α, and likewise B_i in the domain β and C_i in the domain γ, then we say that there are triangles $A_i B_i C_i$ lying arbitrarily near to the triangle ABC.

Axiom III   *If there are displacements of such a kind that triangles arbitrarily near the triangle ABC can be brought arbitrarily near to the triangle A′B′C′, then there always exists a displacement by which the triangle ABC goes over exactly into the triangle A′B′C′.*[27]

The content of this axiom can be briefly expressed as follows:

Axiom III   The displacements form a closed system. We call special attention to the following particular cases of axiom III.

If there are rotations about a point M of the kind that segments lying arbitrarily near the segment AB can be brought arbitrarily near the segment $A′B′$, then there is always such a rotation about M possible by which the segment AB goes over exactly into the segment $A′B′$.

If there are displacements of the kind that segments arbitrarily near the segment AB can be brought arbitrarily near to the segment $A′B′$, then there is always a displacement possible by which the segment AB goes over exactly into the segment $A′B′$.

If there are rotations about the point M of the kind that points arbitrarily near the point A can be brought arbitrarily near the point $A′$, then there is always such a rotation about M possible by which A goes over exactly into the point $A′$.

I now prove the following proposition:

A geometry in which axioms I–III are fulfilled is either the euclidean or the bolyai-lobatchefskian geometry.

If we wish to obtain only the euclidean geometry, it is necessary merely to make in connection with axiom I the additional statement that the groups of displacements shall possess an invariant sub-group. This additional statement takes the place of the axioms of parallels.

In what follows, I will briefly outline the general idea of my method of proof.[28] Within the domain of a certain point *M* construct in a particular manner a certain point-configuration *kk*, and upon this configuration construct a certain point *K*. We then base our investigation upon the true circle *k* about *M* and passing through *K*. It may be easily shown that the true circle *k* is an assemblage of points which is closed and in itself dense. It constitutes, therefore, a perfect assemblage of points.

The next objective point in our demonstration is to show that the true circle *k* is a closed Jordan curve. We do this in that we first show the possibility of a cyclical arrangement of the points of the true circle k, from which it follows that we may map in a reversible, single-valued manner the points of k upon the points of an ordinary circle. Finally, we show that this map must necessarily be a continuous one. Furthermore, it follows also that the originally constructed point-configuration *kk* is identical with the true circle *k*. Moreover, the law holds that each true circle inside of *k* is likewise a closed Jordan curve.

We turn now to the investigation of the group of all the displacements which by the rotation of the plane about *M* transforms a definite true circle *k* into itself. This group possesses the following properties: (1) Every displacement which leaves one point of *k* undisturbed, leaves all points of *k* undisturbed. (2) There always exists a displacement which changes any given point of *k* into any other given point of *k*. (3) The group of displacements is a continuous one. These three properties determine completely the construction of the group of transformations of all the displacements of the true circle into itself. We set up the following proposition: The group of all the displacements of the true circle into itself, which are rotations about M, is holoedric, isomorphic with the group of ordinary rotations of the ordinary circle into itself.

Moreover, we investigate the group of displacements of all the points of our plane by a rotation about M. The law holds that, aside from the identity, there is no rotation of the plane about M which leaves every point of the true circle undisturbed. We now see that every true circle is a Jordan curve and deduce formulæ for the transformation of that group of all the rotations. Finally, the proposition easily follows that: If any two points remain fixed by a displacement of the plane, then all points remain fixed; that is to say, the displacement is the identity. Each point of the plane may be indeed made to go over into any other point of the plane by means of a displacement.

Our further important objective point is to define the idea of the true straight line in our geometry and deduce those properties of it which are necessary in the further development of geometry. First of all, the notions "semi-rotation" and "middle of a segment" are defined. A segment has at most one middle, and, when we know the middle of one segment, then every smaller segment possesses a middle.

In order to pass judgment as to the position of the middle of a segment, we need particular propositions concerning true circles which are mutually tangent, and indeed the question depends upon the construction of two congruent circles tangent to each other externally in one and only one point. We derive also a more general proposition concerning circles which are tangent to each other internally and consequently a theorem covering the special case where the circle which is tangent internally to a second passes through the centre of that circle.

Moreover, a sufficiently small definite segment is taken as a unit segment, and from this by continued bisection and semi-rotation a system of points is constructed of the kind that to each point of this system a definite number a corresponds, which is rational and has as denominator some power of 2. By setting up a law concerning this correspondence, the points of the above system are so arranged that the above laws concerning mutually tangent circles are valid. It is now shown that the points corresponding to the numbers $\frac{1}{2}, \frac{1}{4}, \frac{1}{8}, \cdots$ converge toward the point O. This result is generalized step by step until it is finally shown that every series of points of our system converges, so soon as the corresponding series of numbers converges.

From what has been said, the definition of the true straight line follows as a system of points which arise from two fundamental points, if we apply repeatedly a semi-rotation, take the middle point, and add to the assemblage the points of condensation of the system of points which arises. We can then prove that the true straight line is a continuous curve, possessing no double points and having with any other true straight line at most one point in common. Furthermore, it can be shown that the true straight line cuts each circle drawn about one of its points, and from this it follows that any two arbitrary points of the plane can always be joined by a true straight line. We see also that in our geometry the laws of congruence hold, by which however two triangles are proven to be congruent if they are traversed in the same sense.

With regard to the position of the systems of all the true straight lines with respect to one another, there are two cases to distinguish, according as the axiom of parallels holds, or through each point there exists two straight lines which separate the straight lines which cut the given straight line from those which do not cut it. In the first case we have the euclidean and in the second the bolyai-lobatschefskian geometry.

ENDNOTES

1. Compare the comprehensive and explanatory report of G. Veronese, *Grundzuge der Geometrie*, German translation by A. Schepp, Leipzig, 1894 (Appendix). See also F. Klein, "Zur ersten Verteilung des Lobatschefskiy-Preises," *Math. Ann.*, Vol. 50.

2. These axioms were *neuere Geometrie*, first studied in detail by M. Pasch in his *Vorlesungen uber* Leipsic, 1882. Axiom II, 5 is in particular due to him.

3. Added by Professor Hilbert in the French translation.—*Tr.*

4. See Hilbert, "Ueber den Zahlenbegriff," *Berichte der deutschen Mathematiker-Vereinigung*, 1900.

5. The mutual independence of Hilbert's system of axioms has also been discussed recently by Schur and Moore. Schur's paper, entitled "Ueber die Grundlagen der Geometrie" appeared in *Math. Annalem*, Vol. 55, p. 265, and that of Moore, "On the Projective Axioms of Geometry," is to be found in the Jan. (1902) number of the *Transactions of the Amer. Math. Society—Tr.*

6. See my lectures upon Euclidean Geometry, winter semester of 1898–1899, which were reported by Dr. Von Schaper and manifolded for the members of the class.

7. In his very scholarly book,—*Grundzuge der Geometrie*, German translation by A. Schepp, Leipzig, 1894,—G. Veronese has also attempted the construction of a geometry independent of the axiom of Archimedes.

8. See also Schur, *Math. Annalen*, Vol. 55, p. 265.—*Tr.*

9. F. Schur has published in the *Math. Ann.*, Vol. 51, a very interesting proof of the theorem of Pascal, based upon the axioms I–II, IV.

10. In connection with the theory of areas, we desire to call attention to the following works of M. Gerard: These de *Doctorat sur la geometrie non euclidienne* (1892) and *Geometrie plane* (Paris, 1898). M. Gerard has developed a theory concerning the measurement of polygons analogous to that presented in § 20 of the present work. The difference is that M. Gerard makes use of parallel transversals, while I use transversals emanating from the vertex. The reader should also compare the following works of F. Schur, where he will find a similar development: *Sitzungsberichte der Dorpater Naturf. Ges.*, 1892, and *Lehrbuch der analytischen Geometrie*, Leipzig, 1898 (introduction). Finally, let me refer to an article by O. Stolz in *Monatshefte fur Math, und Phys.*, 1894. (Note by Professor Hilbert in French ed.)

M. Gerard has also treated the subject of areas in various ways in the following journals: *Bulletin de Math*, spciales (May, 1895), *Bulletin de la Societe mathematique de France* (Dec., 1895), *Bulletin Math, elementaires* (January, 1896, June, 1897, June, 1898). (Note in French ed.)

11. *Sitzungsberichte der Dorpater Naturf. Ges.* 1892.

12. *Grundlagen der Geometrie*, Vol. 2, Chapter 5, § 5, 1898.

13. *Monatshefte fur Math, und Phys.* 1894.

14. See also a recent paper by F. R. Moulton on "Simple Non-desarguesian Geometry," *Transactions of the Amer. Math. Soc.*, April, 1902.—*Tr.*

15. Discussed also by Moore in a paper before the Am. Math. Soc., Jan, 1902. See *Trans. Am. Math. Soc.—Tr.*

16. Figures 46, 47, and 48 were designed by Dr. Von Schaper, as have also the details of the demonstrations relating to these figures.

17. "Ueber die Theorie der relativquadratischen Zahlkorper," *Jahresbericht der Deutschen Math. Vereinigung*, Vol. 6, 1899, and *Math. Annalen*, Vol. 51. See, also, "Ueber die Theorie der relativ-Abelschen Zahlkorper" *Nachr. der K. Ges. der Wiss. zu Gottingen*, 1898.

18. See "Ueber ternare definite Formen," *Acta mathematica*, Vol. 17.

19. *Math. Annalen*, Vol. 53 (1900).

20. The following is a summary of a paper by Professor Hilbert which is soon to appear in full in the *Math. Annalen.—Tr.*

21. See Lie-Engel, *Theorie der Transformationsgruppen*, Vol. 3, Chapter 5.

22. By the following investigation is answered also, as I believe, a general question concerning the theory of groups, which I proposed in my address on "MathematischeProbleme," *Gottinger Nachrichten*, 1900, p. 17.

23. Concerning the broader statement of the conception of the plane see my note, "Ueber die Grundlagen der Geometrie," *Gottinger Nachrichten*, 1901.

24. *Abbilden.*

25. Lie makes this assumption to contain the condition that the group of displacements be generated by infinitesimal transformations. The opposite assumption would assist essentially the demonstration in so far as the "true straight

line" could then be defined as the locus of those points which remain unchanged by a displacement changing the sense in which the curve is traversed (*Umklappung*).

26. The term "rotation" is used here in the sense of a rotatory displacement; that is to say, only the initial and final stages and not the aggregate of the intermediate stages of the transition enter into consideration.—*Tr*.

27. It is sufficient to assume that axiom III holds for sufficiently small domains as Lie has done. My method of proof may be so changed as to make use of only this narrower assumption.

28. The complete proof will appear later in the *Math. Annalen*.

Made in the USA
Monee, IL
07 July 2026

56552364R00100